C
THE STATE

Singapore's Built Environment Revisited

Editors

Ooi Giok Ling
Kenson Kwok

IPS
THE INSTITUTE OF POLICY STUDIES

SINGAPORE
OXFORD UNIVERSITY PRESS
1997

Oxford Universtiy Press

Oxford New York
Athens Auckland Bangkok Bombay
Calcutta Cape Town Dar es Salaam Delhi
Florence Hong Kong Istanbul Karachi
Kuala Lumpur Madras Madrid Melbourne
Mexico City Nairobi Paris Singapore
Taipei Tokyo Toronto
and associated companies in
Berlin Ibadan

Oxford is a trade mark of Oxford University Press

First published 1997

ISBN 0 19 588263 6

Printed by Seng Lee Press Pte. Ltd.,
Published by Oxford University Press Pte. Ltd.,
37 Jalan Pemimpin, #03-03 Union Industrial Building, Block A, Singapore 577177

Acknowledgements

We would like to thank Ms Shirley Lim, Administration Manager at the Institute of Policy Studies for her patience and efficiency in the liaison work done between ourselves, the authors and the publishers. Colleagues, particularly, Miss Angeline Yee, have also been extremely supportive. Special thanks to Mr Mok Wei Wei, Mr Lau Woh Cheong and Mrs Lai Choo Malone for their comments on and assistance with the drafts of some chapters in this book and Mr Tay Bak Koi, for the use of his painting of the *Singapore River* on the cover of the book.

We are grateful to Professor Tommy Koh, Director of the Institute of Policy Studies and Deputy Director, Dr Lee Tsao Yuan for their support of this book project.

Grateful acknowledgement is made to all the authors and publishers of extract material which appear in this book:

For Chapters 1, 3, 5, 7, 8 and 10:
Solidarity, Special Issue, Nos. 131/132, July – December, 1991.

For Chapter 9:
Lai, A.E. (1995) *Meanings of Multiethnicity — A Case Study of Ethnicity and Ethnic Relations in Singapore* (Kuala Lumpur: Oxford University Press).

CONTENTS

INTRODUCTION:

Planning Singapore

Ooi Giok Ling and Kenson Kwok

The beginning of this decade promised to be significant to both the city-state of Singapore and Singaporeans. In 1990, Mr Lee Kuan Yew, stepped down as Prime Minister and the younger generation of the PAP (People's Action Party) leaders took over the reigns of political power. While changes of cataclysmic proportions are not expected since this new generation of leaders has been under "tutelage" from the previous generation of politicians for the last decade or so, the Prime Minister, Mr Goh Chok Tong, has held out the promise of a more participatory style of government. Such a promise reflects a departure from past practices, particularly in the area of planning for and the development of the city-state's built environment.

In the past, the government and its agencies have played a crucial and virtually singular role in the transformation of the physical environment in Singapore. This transformation has been unequalled in magnitude and in pace by any previous period in the entire history of this young nation-state (Wong 1989).

The government, headed by the new Prime Minister, has declared its goals for what it terms the next lap in Singapore's physical, economic and social development (Government of Singapore 1991). This is an opportune time to review the past, take stock of the present and also deliberate on future possibilities and developments. Such is one aim of this book which is focused on Singapore's built environment. The bigger objective is to examine the "construction" of the urban community's needs by not only academics but practising professionals and the policy-makers themselves.

When discussing the conceptual problems of urban planning, Harvey (1973, p. 45) had observed that many of the early physical planners possessed a very naive brand of spatial environmental determinism in which a few re-housing projects, a few parks and the like were seen as adequate solutions to the many complex social ills in cities. Yet, Harvey (1973, p. 44) recognised that spatial environmental determinism presented a working hypothesis for physical planners who sought to promote a new social order by manipulating the spatial environment of the city. This approach became a convenient way to break into some of the complexities of the interaction between spatial form and social process for it sets up a simple causal framework in which spatial form affects social process. The approach has been criticised, however, and rejected by sociologists who argue for the democratic notion that what people want is important and the alternative working hypothesis that a social process has its own dynamics so that it will achieve its own appropriate spatial form in spite of the planner (Gans 1969; Sommer 1969; Webber 1963; Jacobs 1961).

Indeed, Webber (1963) has argued that a new spatial order emerges in response to changing technology and changing social norms such that the planner will not be able to prevent the emergence of this spatial order. The spatial determinist, according to Harvey (1973, p. 45), will know that alteration of the spatial structure of the transport network will have repercussions on the social process that will, in turn, generate quite substantial land-use changes. Likewise, the social determinist must realise that if the social process is moving towards some dominant norm such as communication by automobile, then the creation of a spatial form suited for that norm can only reinforce it. Most modern American cities are not built for walking in, and so reinforce the need for car ownership and use.

The observation that conceptual tools for bridging the differences in perspectives between social scientists, geographers and planners have to be sharpened appears to be as much in need as the building of this interface among urban analysts. (Harvey 1973, p. 44). This book is aimed at understanding the social process implications arising from spatial form organisation in a city-state where planning has been highly centralised and the role of the state dominant in the scheme of things. In their effort to evaluate state policies and programmes, the authors have analysed the "construction" of an urban population's needs by the state and its planning organs and matched these against the day-to-day "reality" and experience of the urban populace.

The selection of subjects discussed in the papers that follow is aimed at highlighting the range of social implications arising from the planning and development of the urban built environment. Apart from the issue of safety, culture, ethnic differences, income disparities, private sector participation and the role of public housing residents in the management of their estates represent just some of the social "arenas" for discussing the politics, economics and planning norms that have been applied in the physical development of a city. While the social issues are addressed in the discourse on the planning of the built environment in Singapore, they are not unique to Singapore. Indeed, the social themes underlie on-going concerns being raised in most cities in both the developed and developing world.

In the past, Singapore's policy-makers have focused on pragmatist solutions to the numerous physical problems the city was facing in the late 1950s. These solutions promised expedient and speedy delivery of urban goods and services in sufficient quantities to meet basic needs in housing, sanitation and sewerage, transport and other facilities. The city was faced with so many physical problems, there appeared to be room for improvement in practically every direction one looks and this is evident in the following review by Sandhu and Wheatley (1989, p. v):

> The concentration of at least a third of its 1.5 million polyglot, but ethnically segregated, population in four square miles in the south of the island had resulted in a severe housing shortage that was cxacerbated by widespread decay of the housing stock, fragmented land ownership that inhibited planning and development, and unrestrained population growth. Fewer than one in five dwellings was furnished with sewer-connected sanitary facilities. Industry was poorly developed and its expansion hindered by, among other things, a public transport system that was inefficient by any standard other than profitability and controlled by a management obdurately unresponsive to the needs of the populace.

The challenge to urban policy-makers then was to establish the basis for the building of a modern city-state which could more adequately support the government's plans for social and economic development and growth. In the last three decades, much success has been achieved in modernising the physical environment in Singapore. The focus of urban planning strategies in the past has been on the decentralisation of major land-uses and activities to other parts of the island-state. Action was centred on providing alternative accommodation for people and businesses being relocated from the central area in order to make way for its redevelopment.

In the decades to come, there will be increasingly fewer opportunities for urban expansion on rural land. With a successful public housing programme, industrial estates, schools and excellent infrastructure in place, the challenge being faced by urban policy decision-makers is sustainable urban development. This is not only in the scene of reconciling economic development with limited natural resources and environmental needs but also making Singapore equally if not more, attractive than other cities as a place to visit and do business as well as to live and work in. The city-state's planners recognised that prerogatives of the 1960s and now the 1990s, have changed.

Future demands on the built environment will not be the same as those which have been made before. "We can go beyond the basic necessities of life", observed Mr Ngiam Tong Dow, the then Permanent Secretary to the Ministry of National Development, and "devote some money to quality living" (*Straits Times*, 8 July 1988). The question in the remaining half of this decade and indeed, the century to come, is whether the city planners' view of a "city of excellence" squares with the needs of the urban community, even as these change. A city of excellence, as the political leadership has defined it, is "a gleaming city by day, a glittering wonderland by night," according to Victor Savage who analyses in this book, the translation of the garden-city concept into reality. Savage cites the Prime Minister's intent in his 1984 City of Excellence speech when he was the First Deputy Prime Minister, "We will build a city of expressways and free-flowing traffic. Of time-saving Mass Rapid Transit, cool and comfortable. Of electronics, telecommunications and computers" (*Straits Times*, 12 December 1984).

Based on their studies of the impact of the policies and programmes which have been introduced in the planning, development and management of the built environment, authors of the papers in this book suggest that besides the urban infrastructure, there are other important dimensions to living, working and doing business in the city. The papers emphasise that in striving to be a global city and a city of excellence, attention should be directed to a wider area of concerns arising among the urban community.

The concern for the environment and being green must affect urban policies all the more in the future. Singapore worked at becoming a garden city and the city-state possesses a high standard of environmental quality although the observation has been made that more could be done in energy saving (Friedman 1991). With limited land supply increasingly a constraint on urban growth, nature conservationists are working ever harder to keep our nature reserves intact.

The aspiration to be a good citizen in the global community and the emphasis on an international outlook means that Singapore must be conscious of the world community's concern with population growth policies and their demands on natural resources. Policy-makers in Singapore are aware of the need to address the growing proportion of the ageing in the population. They have been concerned with accommodating the needs of a multi-cultural and multi-ethnic society. Increasingly, there is recognition that more options can be provided for urban action by involving the private sector in urban policy decision-making.

A more participatory style of government appears to be appropriate for the decade to come. In a survey at the beginning of this decade, Singaporeans were of the view that they were over-regulated (Institute of Policy Studies 1990). Official channels that have been instituted to provide Singaporeans with more opportunities to speak their minds on issues affecting them have been widely used. Institutional changes such as the introduction of town councils to manage public housing estates appear to have been in the same spirit of involving the public more in decision-making on the built environment (Ooi 1990). While policy-making in the past has been highly centralised and never pluralistic in nature (Lee 1989), there is recognition that, in recent years, there has been some re-politicisation of the citizenry (Chan 1989). This has spawned the view that a compromise will ultimately be reached. For the leadership, the compromise will be somewhere between economic effectiveness and political stability and for the citizenry between economic well-being and increased political participation (Chan 1989).

The built environment is the most visible testimony of Singapore's rapid development into one of the four "little dragons" of Asia. This city-state is also the most highly urbanised of the ASEAN (Association of Southeast Asian Nations) member countries. The papers assembled in this book discuss the tremendous physical changes which it has experienced since gaining independence from British colonial rule in the late 1950s. These papers also deal with the impact and implications arising from the changes before tackling the more demanding task of suggesting future orientations and considering new directions.

In its aspiration to become a global city, the city-state of Singapore has, in the coming years, to consider global standards by which it would be judged anyway. It had the world's number one airport for the years 1988 to 1990 and the busiest port for the fifth consecutive year in 1990 (Wu and Ng 1991). In the coming century, Singapore, as a city, will continue to tackle issues concerning environmental standards, urban transport and land-use while striving to accommodate its population's demands for a better quality of life. The issues related to the development of the built environment will change and hence, assumptions upon which policies have been formulated even as socio-economic conditions of the population and their needs change. For the professed purpose of developing Singapore into the people's city, urban policy-makers have to be conscious of such demands on the built environment and adapt planning policies and practices accordingly. This book makes a start towards the evaluation of some of the issues arising from the development direction which Singapore's built environment has taken and the people's perception of the city today.

Evaluation is done and urban planning and development theories considered. Urban transport planners have, for example, to consider the contention that private modes of transport and hence, public road development will never be able to accommodate the needs of a growing urban population (Castells 1977). A balance has, therefore, to be struck between public and private means of transport. This balance between public and private sector contributions applies equally to other areas in the development of the built environment. There needs to be a recognition of the significance of market forces and the role of the private sector in policy decision-making on the built environment (Lim et al. 1989).

The papers in this book can be grouped into five categories. Planning the built environment is the theme of the first category of papers. Singapore's new chapter in physical development as well as the results of past planning experience form the focus. What form is the built environment likely to take in the future? In her paper, **Sim Loo Lee** describes a scenario which features marinas and luxury apartments in the central area together with the conservation of old buildings, particularly in ethnic neighbourhoods and the upgrading of public housing estates which are largely located outside of the central area. In this upgrading exercise which she deems more difficult than the preceding phase of new public housing development, she foresees the need for less uniformity and greater visual identity in the public housing estates of the future.

Liu Thai Ker, former Chief Executive Officer of the Urban Redevelopment Authority and Chief Planner, takes up the theme of the need to introduce greater variety in the physical development of the city-state. In his paper on Singapore as a tropical city of excellence, he sees more effort being put into creating greater variety in development and environmental characteristics while maintaining high environmental quality.

Ooi Giok Ling discusses another area of change — the introduction of town councils in Singapore to manage public housing estates formerly, entirely dealt with by the public housing government authority. Town councils portend the introduction of a new social and administrative order to complement what has become a spatial order which has evolved from the re-organisation of land uses through planning and development by the public sector agencies.

Phang Sock Yong contributes another perspective on the impact of the spatial order on urban mobility, that is, the city's transport needs. In the wake of the newly created Land Transport Authority's blue print for a world-class transport system for Singapore, Phang's assessment of the need for greater and closer integration between land use and transport is a useful starting point to contemplate future directions for transport planning and development. The paper focuses not only on the nuts and bolts of the urban transport system but also on the impact of policies related to public housing and such urban land uses.

The second category of papers concentrates on the economics of the built environment. Two papers, the first on the political economy dimensions of the built environment and the second; on business and the built environment, deal with roles of the public sector, market forces and the private sector in the physical development of Singapore. **Linda Low** argues in her paper that the private sector should play a greater role in the development of Singapore's built environment. She points out the leading role of the public sector as a major land-owner and also in shaping the built environment in the last 25 years.

The argument for greater participation, particularly, by the private sector is pursued further in **Ooi Giok Ling's** paper. In this paper, the author has stressed the necessity for greater co-operation between public and private sectors, particularly in planning and development to meet the future needs of the business community with the goal of increasing the range of choices and options for business. This is seen as crucial to the quality of the built environment and its competitiveness *vis-à-vis* other cities in terms of its attractiveness as a business centre.

With practically nine in ten persons living in public housing, the third category of essays has focused on the socio-cultural dimension and concerns with the public housing environment. **Sharon Siddique** looks at how the perceptions and behaviour of different ethnic groups have been shaped by the public housing environment. While it has been the policy to have ethnically integrated public housing estates, she is concerned, in her paper, with the extent to which those with the power over the built environment — policy-makers and estate managers diverge from the cultural "traditions" of people who live in public housing.

A similar concern is reflected in **Lai Ah Eng's** paper on inter-ethnic relations among public housing residents and the role of housing estate managers. In her paper, she discusses the coping strategies employed by the public housing residents in their everyday and routine dealing with neighbours from different ethnic backgrounds as well as the contributions of housing and maintenance officers to the settlements of disputes. In both their papers, the two authors demonstrate the importance of official understanding and support in accommodating ethno-cultural sensitivities and maintaining good inter-ethnic relations.

Similarly, **Myrna Blake** emphasises greater consciousness on the part of public housing officials as yet another aspect of public housing development. She evaluates the built environment in relation to the needs of the underpriviledged such as the poor, the illiterate and the elderly. In their transactions with the Housing and Development Board (HDB), the underprivileged have faced severe constraints arising from the bureaucracy's single-minded provision for the "average" Singaporean family. She recommends various procedures which, when implemented, would mean the better accommodation of the underprivileged in future.

In the fourth category of papers, the concern is with the public perception of various features of the built environment. **Victor Savage's** paper evaluates the concept of the garden city as it has been interpreted in Singapore and manifest in the built environment or cityscape. He examines the ideological overtones of the garden city and suggests ways in which the concept and its future implementation could be made more coherent and meaningful for the people living in or visiting the city.

Finally, there are two essays on first, safety of the built environment and then, the sense of security among public housing residents. This is the International Decade for Natural Disaster Reduction for good reason. According to Whittow (1995) and Degg (1992), the 1990s have been declared the decade for natural disaster reduction because of a dramatic increase in the number of recorded disasters such as floods, earthquakes, and volcanoes. The January 1995 Kobe earthquake disaster has been a strong reminder that in heavily urbanised areas and highly populated cities in earthquake prone nations such as Japan, the built environment remains at considerable risk in spite of major technological advances recently achieved in earthquake engineering.

In prescribing measures to be taken to mitigate the impact of natural disasters, professionals and other researchers have different views. Civil engineers, while agreeing with the need for programmes and planning for aid and support in the aftermath of a disaster, also emphasise preventive measures in addition to post-disaster relief and rehabilitation schemes. This discussion is addressed in the paper by **Kog Yue Choong** and **T. Balendra**. Their paper focuses on the safety of high-rise environment which has been such a prominent feature of Singapore's physical development in the last three decades. In their analysis of the regional location of Singapore, local soil and geological conditions and their relationship with the built environment, the two authors argue the case for the future need to monitor the impact of earth tremors in the region on the city's buildings. The significance of their argument should be enhanced by the growth in high-rise development in the future.

The paper on the fear of crime and sense of security in public housing estates by **Brenda S.A. Yeoh** and **Yeow Pei Lin** considers the public housing residents' assessment of how safe they feel in their estates. While noting that the crime rate is low in Singapore and the streets safe to walk at all times of the day, the authors, nonetheless, suggest that further consideration could be given to improving the sense of safety among residents when they move about their estates. There is a focus on women's concerns about the features of the built environment and the landscaping works which contribute to making them feel unsafe when moving about alone in their neighbourhood. The authors argue that the public housing authority and designers in particular, should give more thought and take the necessary steps towards enhancing the sense of safety in public housing estates. Some emphasis has been placed on common areas in these estates. If these areas cannot be considered as "no man's land" because of the fear of crime and indeed, are far from being so, the authors provide evidence that more needs to be done so that the common areas in public housing estates do not become "no woman's land" for fear of crimes against women such as molest.

Singapore stands poised at the brink of a new phase in the development of its built environment. People who are involved in the shaping of the built environment must contend with a range of issues facing Singapore's future development. Without having exhausted them, the papers on Singapore's built environment which are presented in this book have tackled a variety of the issues involved. Effort that is directed at addressing these issues can only augur substantive improvement for the quality of the built environment as well as make it more accommodative of the varying needs of the city and its people. These improvements will contribute towards the aim of making the city the place it aspires to be in the coming century — a city of excellence and more importantly, the people's city.

REFERENCES

Castells, M. (1977) "The Urban Question", (London: Edward Arnold).

Chan, H.C. (1989) "The PAP and the Structuring of the Political System", in K.S. Sandhu and P. Wheatley (eds.) *Management of Success — The Moulding of Modern Singapore* (Singapore: Institute of Southeast Asian Studies), pp. 70–89.

Degg, M.R. (1992) "Natural Disasters — Recent Trends and Future Prospects", *Geography*, Vol. 77, No. 3, pp. 198–209.

Friedmann, J. (1991) "The Making of World Cities", Paper presented at the Seminar on World Cities", National Productivity Board, 23 March, Singapore.

Gans, H.J. (1969) "Planning for People Not Buildings", *Environment and Planning I*, pp. 33–46.

Government of Singapore (1991) *The Next Lap* (Singapore: Times Editions).

Harvey, D. (1973) *Social Justice and the City* (Baltimore: Johns Hopkins University Press).

Institute of Policy Studies (1990) Study of "The Singaporean — Ethnicity, National Identity and Citizenship", Unpublished report, (Singapore).

Jacobs, J. (1961) *The Death and Life of Great American Cities* (New York: Random House).

Lee B.H. (1989) "The Bureaucracy", in K.S. Sandhu and P. Wheatley (eds.) *Management of Success — The Moulding of Modern Singapore* (Singapore: Institute of Southeast Asian Studies), pp. 90–101.

Lim, C.Y. et al. (1988) *Policy Options* (Singapore: McGraw-Hill).

Ooi, G.L. (1990) "Town Councils in Singapore — Self-Determination for Public Housing Estates", Occasional Paper

No. 4, Times Academic Press for Institute of Policy Studies, (Singapore).

Sandhu, K.S. and P. Wheatley (eds.) (1989) *Management of Success — The Moulding of Modern Singapore* (Singapore: Institute of Southeast Asian Studies).

Sommer, R. (1969) *Personal Space — the Behavioural Basis of Design* (Englewood Cliffs, New Jersey: Prentice Hall).

Webber, M. (1963) "Order in Diversity: Community Without Propinquity", in Wingo, L. (ed.) *Cities and Space: the Future Use of Urban Land* (Baltimore: John Hopkins University Press).

Whittow, J.B. (1995) "Disaster Impact and the Built Environment", *Built Environment*, Vol. 21, No. 2/3, pp. 81–88.

Wong, P.P. (1989) "The Transformation of the Physical Environment", in K.S. Sandhu and P. Wheatley (eds.) *Management of Success — The Moulding of Modern Singapore* (Singapore: Institute of Southeast Asian Studies), pp. 771–781.

Wu, F. and Ng, B.E. (1991) "Singapore's Strategy for the 1990s", *Long Range Planning*, Vol. 24, No. 2, pp. 9–15.

1

Planning the Built Environment for Now and the 21st Century

Sim Loo Lee

Introduction

The built environment of the city-state of Singapore has undergone dramatic changes over the past three decades. From an entrepot port and colonial city with a landscape that has been dominated by traditional two-storeyed shophouses in the 1960s, it had developed into a metropolis with ultra-modern, high-rise buildings by the end of the 1980s. More changes and development can be expected since the government has announced its plan to make Singapore the first developed city in the equatorial belt by the turn of the century.

Since the mid-1980s, many physical development plans have been formulated to transform the city. These plans show a shift in policy from a quantitative approach to one of achieving quality and excellence in the built environment. However, the plans have been undertaken on a rather piece-meal basis by different government planning agencies such as the Urban Redevelopment Authority (URA), the Housing and Development Board (HDB), the Jurong Town Corporation and others. The plans are, therefore, lacking in co-ordination.

Not surprisingly, in 1989, substantial changes were introduced in the planning organisation and hence, the town planning process. Government agencies in charge of planning were reorganised. The Planning Department and the Research and Statistics Unit of the Ministry of National Development were merged with URA. With this reorganisation, the URA is now the planning authority in charge of all planning functions for the physical development of Singapore, including the conservation of land and buildings.

Where the town planning process is concerned, 1989 saw a major policy change. The private sector was invited to prepare the Development Guide Plans (DGPs) for two areas — Simpang and Kampong Bugis. When completed, these plans were reviewed alongside those prepared by government planners before the plans were finalised. In 1991, the private sector was invited to prepare the DGPs for Changi and Jurong East, and in 1993, the DGPs for the Nanyang region around the Nanyang Technological University and Geylang East. By involving the private sector, the government is hoping to obtain fresh perspectives as well as new ideas and concepts to achieve improved quality in the built environment.

A public consultation process was introduced in 1988 through which the private sector has been invited to comment and participate in the planning process through dialogue sessions with government planners. The private sector has, so far, participated in dialogue sessions concerning the draft Civic and Cultural Master Plan, the draft Master Plan for the Urban Waterfronts at Marina Bay and Kallang Basin, and the draft manuals for the conservation of Chinatown, Kampong Glam and Little India. When the review of the Concept Plan was completed in 1992, the URA invited the private sector to comment and participate in the planning strategies.

The revised Concept Plan (see Fig.1.1) which will guide the physical development of Singapore right through to the 21st century sets out the strategies for creating a tropical city of excellence. These strategies include the conservation of historically and culturally rich areas, the development of a new downtown, decentralisation and the establishment of regional centres, land sales in new towns for more private middle-class housing and more land for recreation. Together, these strategies and the policy for improving public housing to achieve an excellent built environment by the year 2000 are discussed in the following sections. The discussion here is confined to plans for the year 2000 only as these are more likely to be implemented within the next few years compared to plans for the year 2030.

Conservation

The first step towards achieving quality in the built environment is the shift in policy from demolition and redevelopment to conservation in the Central Area for the next decade. The government has often been criticised for destroying historically and culturally rich areas. Conservation, it is hoped, will retain our cultural heritage and vital links to the past. Towards this end, a comprehensive Cultural Master Plan was prepared for the conservation of areas rich in history, culture and architecture.

Prior to the introduction and drawing up of this Cultural Master Plan, conservation efforts were limited to small areas such as Emerald Hill and Peranakan Place. The success of these small projects has probably encouraged the government to embark on conservation on a more comprehensive scale.

Under the Cultural Master Plan, five major areas considered to be rich in history and culture have been identified for conservation. The five areas demarcated are Chinatown (comprising Kreta Ayer, Telok Ayer, Tanjong Pagar and Bukit Pasoh), Kampong Glam, Little India, The Singapore River (Boat Quay and Clarke Quay) and the Civic and Cultural District (see Fig. 1.2).

Chinatown, Little India and Kampong Glam are ethnic districts, historically and culturally associated with the Chinese, Indians and Malays respectively. The objectives of conserving these areas are to restore buildings of historical and architectural significance, retain and enhance ethnic-based activities and improve the general physical environment. Traditional trades and ethnic-based activities will be retained while new, compatible ones will be introduced in the core areas. Further, appropriate new features to enhance the identity of each place will also be incorporated.

Guidelines for the conservation of Chinatown, Kampong Glam and Little India were completed in mid-1988. A pilot conservation project was implemented in Chinatown at the end of the same year. The first phase of 32 shophouses at Tanjong Pagar was sold by tender and refurbishment works have since been completed successfully (Tan 1989). These shophouses are for mixed use — with retail on the ground floor and retail, office and residential uses on the upper floor (see Fig. 1.3). After restoration, their value has more than doubled, negating the argument that conservation of historical areas in the Central Area is a gross under-utilisation of prime land and does not make economic sense.

Another conservation area, the Civic and Cultural District, considered to be the historical and cultural core of Singapore, will be revitalised as a showpiece of colonial history and architecture, and a retail magnet of the city. Distinctive qualities of the Civic District will be accentuated so as to reinforce its identity. The special functions of the District will also be enhanced to make it the venue for national celebrations and ceremonies.

The government has announced that it would take the lead in conservation by renovating state-owned properties, covering approximately 25 per cent of the areas being preserved, within a five-year period. It would also introduce landscaping and provide pedestrian walkways and plazas in these areas. Since 1994, Bussorah Street in Kampong Glam has been turned into a pedestrian mall.

It is heartening to note that conservation plans have not been confined to the preservation of groups of buildings of historical and architectural significance. These plans cover wide areas encompassing mixed uses and activities of different ethnic groups. However, the balance of these elements is crucial for preserving each area's distinctive character and ambience. Such a balance should be struck between the forms of old buildings and new developments, buildings and open spaces, and among different uses.

Moreover, one has to identify clearly the ends, the uses and the benefits. Which group of users will be served? Locals or tourists or both? Opinions on these issues differ, but decisions will have an effect on the details and implementation of specific developments and schemes for the preservation of the character and ambience of these different areas designated in the plans.

Studies on the conservation areas have shown that conservation has been effective in restoring and preserving buildings of historical and architectural value. It has also succeeded in retaining traditional trades and ethnic-based activities in the core areas (Sim, Brown, Tan and Leong 1995). However, outside the core areas uses are left to market forces. An economic evaluation of the Tanjong Pagar Conservation Scheme by Wong (1990) has shown that businesses are not performing well. Similarly, a study on retailing in the Kreta Ayer area has revealed that traditional trades are fast disappearing from the conservation area (Low 1992) as rental rates of the restored shophouses have increased tremendously.

New Central Business District

Among the many projects being planned, the most interesting for planners, architects, developers, investors, businessmen and in fact, Singaporeans from all walks of life, would be the development of a new city on 690 hectares of reclaimed land (see Fig. 1.4). A new downtown core will be created as a logical extension of the "Golden Shoe", the core of the present Central Business District. The focus of the new "Golden Doughnut" will be Marina Bay, an expanse of water which will be a venue for national events. Buildings with striking appearances will be strategically positioned around the Bay to serve as major landmarks. A special focal point will also be created at the mouth of the Singapore River.

The DGPs for the Downtown Core and Portview were completed and exhibited in August 1992 to obtain feedback from the public. These DGPS provide the details that will transform the vision of the new downtown in the revised Concept Plan into reality. The new downtown was conceived to be an international centre for business and commerce and an entertainment hub.

The downtown core will extend into Marina South. A waterfront promenade will stretch all the way from Marina South to Marina Centre. This will be linked with the open spaces and greenery of the city to further project our Garden-City image, and to provide a haven for office workers, weary shoppers, tourists and casual strollers.

Framing the waterfront promenade will be commercial developments comprising offices, shopping centres and hotels.

To inject life into this new central business district, substantial areas of Marina South and Marina East will be reserved for residential use. Luxury apartments with panoramic views of the Bay will be developed.

Great attention has been paid to the scale and definition of urban forms and spaces, and to activities which will generate interest and liven up the new downtown. Such activities are crucial for a dynamic city centre. In this regard, the recreational and retail facilties such as little restaurants and cafes along the waterfront promenade and the theatre (which is being built) will be crowd-pullers during evening hours, weekends and public holidays.

Plans have also been unveiled to redevelop the Kallang Basin as an extension of Marina Bay. Sites will be put up for sale in the next few years for the development of recreational projects and luxury waterfront apartments with marinas. To optimise the use of this water-body and its beaches, related recreational activities such as boating, sailing, fishing and water-skiing should be encouraged.

The total image of the new downtown will be visually exciting and modern, in keeping with the 21st century. Whether this will be a reality by the year 2000 depends on the joint participation of government agencies and the private sector. Great effort will be required from both sectors to create and organise a functional urban environment where Singaporeans can live, work and play. Judging from the successful joint participation of both sectors in the redevelopment of the Central Area, there is a possibility that the new downtown will be a reality.

Decentralisation

Besides conservation in the Central Area and development of the new downtown, there is also a policy to establish major regional centres to encourage decentralisation from the Central Area (Liu 1990). In this programme, the government intends to take the lead by relocating government and quasi-government activities that do not require a central location. One example would be the development of a building for the Central Provident

Board at Tampines Regional Centre. Incentives are also being planned to encourage commercial activities to relocate outside the Central Area.

Decentralising and relocating commercial activities, especially offices in the regional centres, will ease congestion in the Central Area. However, the choice of location for such centres is crucial for the success of this decentralisation effort. Locations identified for establishing commercial centres are Jurong East, Woodlands, Tampines and Seletar (see Fig. 1.1). These are large population centres and as such, they can be expected to have great potential for success as regional centres.

Other possible locations which can be developed into commercial nodes are selected Mass Rapid Transit (MRT) stations. Suburban MRT stations within the HDB new towns such as Clementi, Bedok and Ang Mo Kio have great potential to be developed into complete townships or sub-regional centres. Thus far, the HDB new towns have been developed as high-density residential satellite suburbs with lower-order retail centres, not full-fledged townships. The government's recent policy of introducing private sector developments in the form of high density middle-income housing into HDB new towns will create the necessary threshold for more comprehensive and higher order commercial developments.

The sub-regional centres, the regional centres and the Central Business District, well-connected by a system of expressways and roads and the MRT, will form a hierarchy of centres serving different regions of the island (see Fig. 1.1). Questions have been asked regarding the new urban structure. The question most frequently raised is whether there is a need for multiple-growth centres within such a small city-state. Given the short distances between the growth centres due to the small size of the island, these centres will eventually coalesce and form an integral part of a continuous urbanised area. So, are regional centres and sub-centres necessary?

Another question is the quantum of commercial space that will be provided in the regional and sub-centres. Will there be a gross oversupply? Can a population of four million sustain the quantum of commercial floor space proposed within the new hierarchical structure? It is good for planners to envisage rapid urban growth and allocate resources to meet the needs accordingly. However, the questions raised have to be considered carefully at the implementation stage.

Land Sales in New Towns

One part of the decentralisation programme involves 13 parcels of land totalling to ten hectares within the HDB towns which will be sold for commercial development. At the end of 1993, several sites for commercial development in new towns for example, Yishun, Tampines, Jurong East, Jurong West, Choa Chu Kang, Bishan and Bukit Merah have been sold by tender. Commercial developments comprising office blocks, shops and cinemas will be developed on these sites by private developers.

With private sector initiatives, it is likely that these town centres will develop distinctive characters and individuality, unlike the town centres created solely by the HDB. In addition to commercial sites, parcels of land in the HDB new towns have been set aside for the private sector to develop middle-income housing. The HDB and the URA have begun to release 17 sites, comprising a total of 43 hectares. Altogether, these will supply 3200 housing units comprising flats, condominiums, townhouses and landed properties. The aim of this policy is to achieve integration of public and private housing developments.

At present, almost all private housing developments are segregated from the public housing estates such that mass public housing blocks with uniform facades have created a monotonous built environment. The integration of public and private middle-income housing will result in different types of housing and hence, different scales of environmental identity. Another benefit will be the integration of different income groups which makes the government's effort to promote community living easier.

The first two sites (one in Tanah Merah and another in Telok Blangah) totalling to some 35 700 square metres were released for sale in early 1990. Some 260 apartments ranging from 120 square metres to 158 square metres, with a total development cost of $100 million were planned for the first site, while the second smaller site consisted of 150 apartments ranging from

112 square metres to 177 square metres, to be developed at a total cost of $70 million. Both projects were completed in early 1994. Following these two, other residential sites at Serangoon and Clementi New Towns were sold by the HDB in December 1990. Condominiums have been completed on these sites since the end of 1994.

Land sales in the HDB new towns for commercial and private middle-income housing will bring about private sector participation within the HDB new towns which, hitherto, have been exclusively, the domain of the public sector. This move is welcomed by private developers who are finding it increasingly difficult to acquire land for development in land-scarce Singapore. It will also increase the supply of affordable private housing for middle-income Singaporeans who wish to upgrade from public HDB flats to private housing.

Public Housing

Where public housing is concerned, the HDB has achieved spectacular success in, and received international recognition for its massive public housing programme (Wong and Yeh 1985). Since its inception in 1960, the HDB has constructed some 650 000 units, housing some 87 per cent of the population.

In keeping with the government's policy of achieving excellence in the built environment by the end of the decade, the HDB is now shifting its emphasis from quantity to quality, that is, from meeting housing needs to satisfying the preferences of the people. Public housing policy for the next decade is to provide bigger and better quality homes to meet the preferences of an increasingly affluent population. Thus, in the revised Concept Plan, provision has been made for each Singaporean to occupy between 30 and 35 square metres compared to the present 20 square metres.

In July 1989, the government announced a $15 billion public housing upgrading exercise which is expected to span two decades and to benefit 95 per cent of HDB dwellers. The first phase, involving pilot projects in Woodlands and Teban Gardens, have been completed and upgrading works have also been completed at six demonstration precincts. The upgrading covers three aspects, namely, improvements inside the flat, to the block and to the estate. The last two aspects will cause substantial changes to the built environment.

The facades of blocks of flats will be given a face-lift, with additional architectural enrichments such that the plain-looking blocks at present will assume more individual character. In the upgrading of the housing estate, the planning unit used will be the precinct, introduced in estates developed during the late 1970s. This concept involves the designing of public housing estates into small precincts, each comprising five to seven apartment blocks with a total of up to 750 apartments. These blocks will share a common open space with landscaped gardens, children's playgrounds, barbeque pits, and pavilions. Each precinct will have its own distinctive character and identity to give residents a greater sense of belonging, privacy and a pleasant living environment similar to those found in private condominiums.

In introducing these innovations, the HDB shows that it is now paying more attention to the micro-structure of the built environment. Undoubtedly, upgrading will get rid of the uniform appearance of public housing estates resulting in a better built environment but it will be an extremely difficult task.

When the programme was announced, there were doubts about whether upgrading on such a massive scale could be achieved. First, initial response by residents was not too enthusiastic because of the requirement that flat-owners pay 30 per cent of the cost incurred. An early survey by *Straits Times* had shown that more than half of the respondents were unwilling to pay. Furthermore, residents had expressed the fear that more facilities would entail higher maintenance or conservancy charges. Then, at least 75 per cent of the residents have to opt for the upgrading before it could be implemented. Finally, most of those reluctant or unable to pay were among the older generation who happened to live in the older estates which were the very ones in need of upgrading. The government, therefore, faced an uphill task.

Notwithstanding all the problems mentioned above, the upgrading programme, at least in its pilot phases, have been successful. Response of the residents in the demonstration precincts has been overwhelming due to, perhaps, great publicity and promotion (Foo 1993). Encouraged by this, the HDB has decided to upgrade six more precincts and introduced an interim upgrading programme alongside the main one.

Private Housing

While the government through the HDB has catered to the housing needs of the majority of the lower-income population living in public housing, that of satisfying the demand of the higher-income group has been left to private developers. Over the past two decades, private housing which caters to the upper income group has been in great demand, due to increasing affluence among Singaporeans.

In an attempt to maximise scarce land resources and to satisfy the great demand for private housing in the 1970s, the government introduced the condominium concept of housing. Since its inception in the late 1970s, condominiums have mushroomed all over suburban Singapore (Sim 1989). Of varied architectural designs, condominiums present a sharp visual contrast to the low-rise conventional types of housing and rather uniform high-rise HDB blocks, enriching the built environment (Teo and Savage 1985). Often well-equipped with a wide range of recreational facilities and beautifully landscaped gardens, they provide residents with a country-club lifestyle and more gracious living.

Although the condominium concept of housing has been successful both in maximising land use and satisfying the demand for private housing, the government has begun to look into alternative forms of private housing for the next decade. The revised Concept Plan aims to provide more opportunities for Singaporeans to enjoy the benefits of medium- and low-density housing. Accordingly, the proportion of medium- and low-density housing will be increased from 17 per cent to 30 per cent. At the same time, Singaporeans will be provided with a greater variety of housing forms.

In order to tap ideas from private architects and developers, a new housing concept competition was launched by the Ministry of National Development in April 1990. Participants were given a choice of three plots of land in Newton Road, University Road and Hillview Avenue to propose new forms of residential developments. They were urged to explore new forms of high-, medium- and low-density developments, a departure from the familiar high-rise apartment blocks.

Among the new concepts that emerged from the competition, cluster housing with recreational facilities similar to those found in condominiums seems to be the attractive alternative. Basically, the concept involves building a cluster of L- or U-shaped houses sharing common facilities such as gardens, a swimming pool and maybe, a tennis court. At the same time, privacy can be retained by using low-garden fences and hedges to divide and screen the units. Another new form of low-density housing suggested by the private sector would be strata bungalows. Both of these new housing forms combine the quality of the environment of landed properties with condominium-style facilities. The URA has decided to adopt these new housing forms and recently, detailed guidelines were made known to the public at a workshop (Han 1993).

Besides the search for new housing forms, the URA has also reviewed planning guidelines for private housing. More flexible standards have been set to encourage greater creativity and to ensure that land is put to optimum use. The new planning guidelines allow for smaller plot sizes for semi-detached houses, less setback from the road, terrace houses to be built directly next to the road and bigger built-in area for bungalows. Condominium guidelines were also revised to allow for greater flexibility in the design and layout of developments (Wong 1993).

New concepts in private housing will, undoubtedly, result in many interesting housing forms such as the U-shaped and L-shaped arrangement for terrace houses. Such housing, arranged in clusters, will also help to break the monotony of rows of conventional terrace houses. These will not only add variety to the existing housing forms but also result in a more interesting skyline. New guidelines will also help to achieve the Concept Plan's objective of providing a wider variety and choice of housing, enriching the built environment in the next decade.

Conclusion

Clearly, the Singapore government is now orientating its approach to satisfying not only physical requirements, but also the psychological, social, emotional and aesthetic needs of the people. In order to achieve this objective, many interesting plans have been formulated to transform the built environment. If successfully implemented, Singapore will definitely possess a high quality built environment by the year 2000. Singaporeans will then have an exciting modern downtown, a historically and culturally rich Central Area, a choice of different housing types with varied densities and more recreational facilities.

However, development of the built environment is a continuous and dynamic process dependent on a country's economic and political health. Singapore has been successful in shaping its built environment over the past three decades because of several factors.

First, the city-state's small territory makes it easier to plan, manage and implement policies; development planning has, since the 1960s been equivalent to national planning (Yeung 1973). Second, it has also been easier to educate and train its small population into a highly skilled workforce, making available professional, managerial and technical personnel. Third, rapid economic growth has generated adequate financial resources for development projects comprising infrastructure, housing and others. Fourth, Singapore's political stability has promoted confidence and attracted huge local and foreign investments in the built environment. Finally, a relatively corruption-free machinery, and a clear system of rules and regulations have helped to expedite the implementation of policies. The government's policy to develop a great city with an excellent built environment in the next decade is highly dependent on the conditions discussed above, and can only succeed if these continue to prevail.

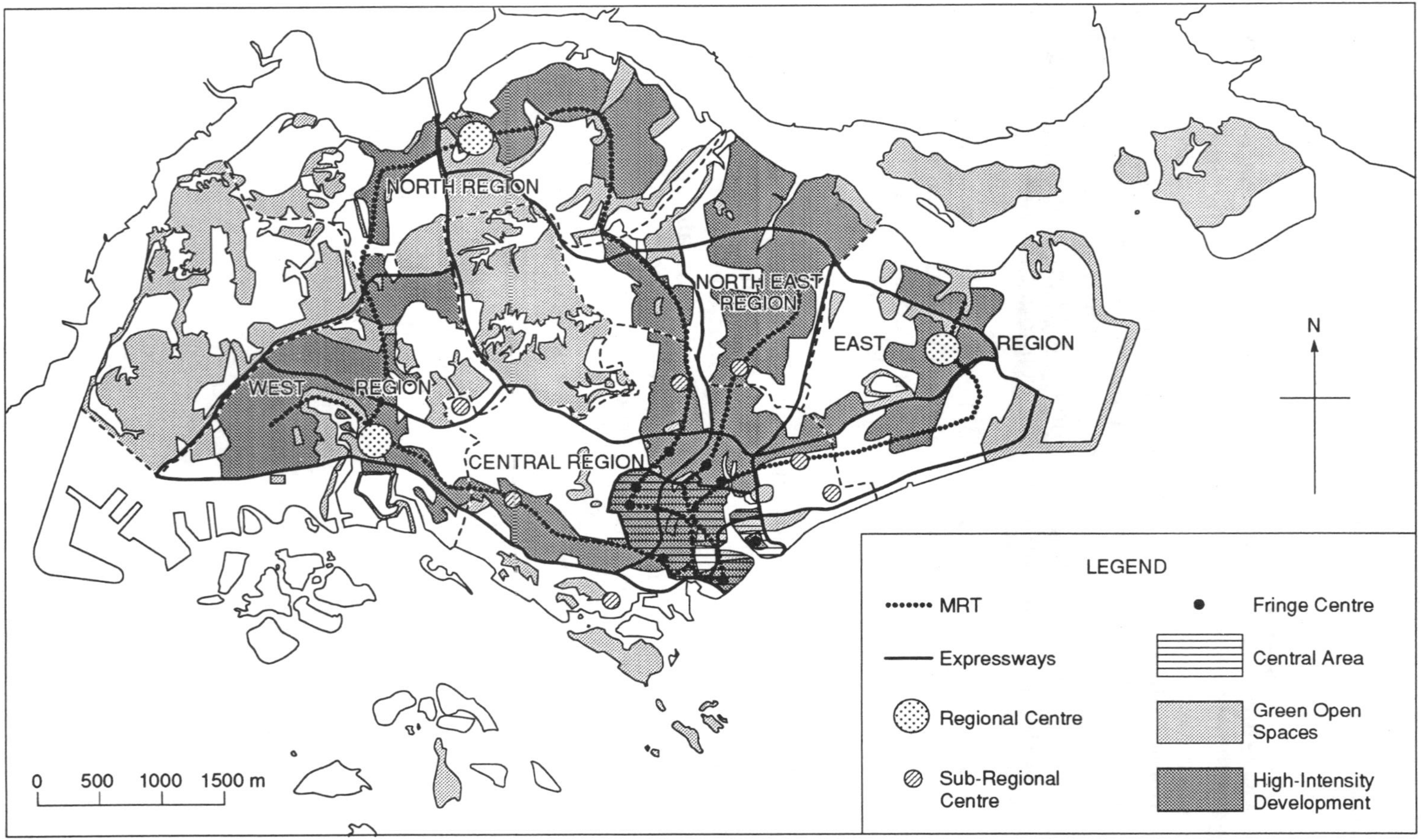

Source: Urban Redevelopment Authority (1988).

Figure 1.1 Singapore Concept Plan

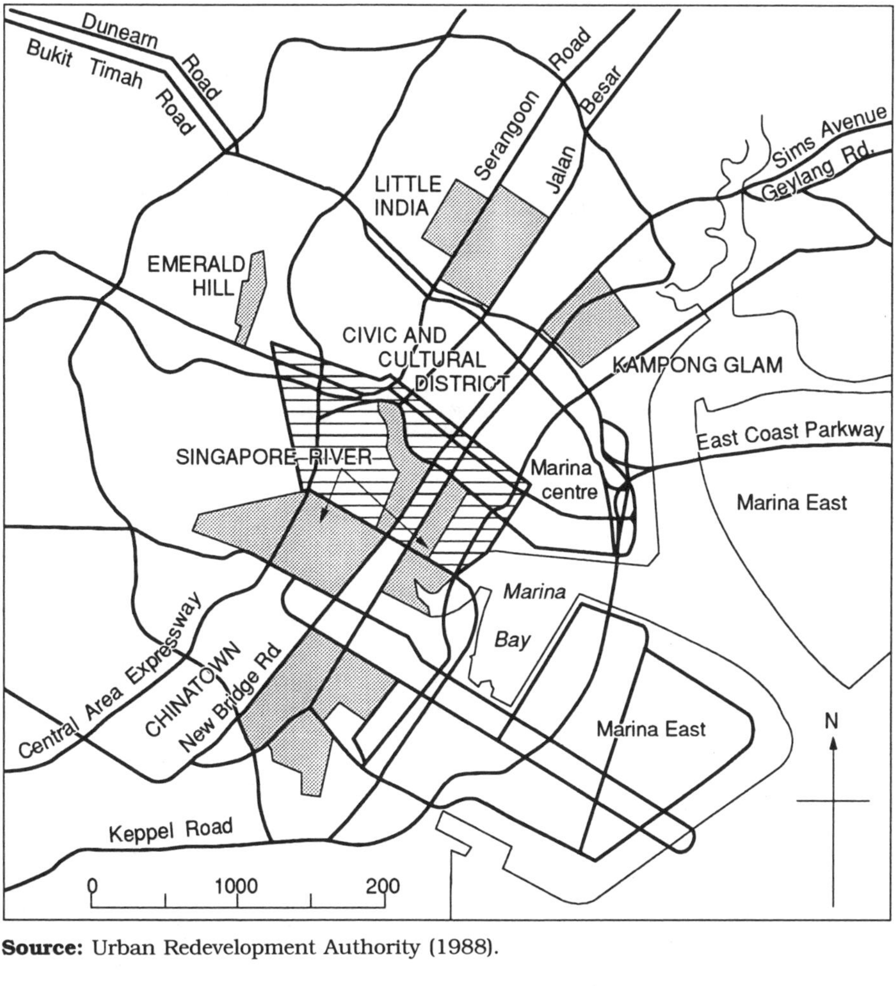

Source: Urban Redevelopment Authority (1988).

Figure 1.2 Historic Districts under Conservation

Figure 1.3 Conservation of Shophouses

Figure 1.4 New Downtown

REFERENCES

Foo, T.S. (1993) "Factors for the Success of Upgrading HDB Flats", Unpublished M. SC dissertation, National University of Singapore, (Singapore).

Han, Y.H. (1993) "Guidelines for Cluster Housing and Strata Bungalow Developments", SIA/SIP Planning Workshop on Strata Bungalows, Cluster Housing and Revised Housing Guidelines, July, Singapore.

Liu, T.K. and Lau W.C. (1979) "Images of the Man-Made Environment towards the Year 2000", Seminar on Singapore towards the Year 2000, 20–21 December, Singapore.

Liu, T.K. (1989) "Improving the Living Environment of Singapore", *Fourth Symposium on Our Environment* (Singapore: Singapore University Press).

Low, K.Y. (1992/1993) *"The Future of Retailing in Kreta Ayer"*, Unpublished dissertation, National University of Singapore, (Singapore).

School of Building & Estate Management (1990) *Focus on Property in Singapore*, National University of Singapore, (Singapore).

Sim, L.L. (1989) *Guide to Condominium Housing in Singapore — Western* (Singapore: Singapore University Press).

Sim, L.L. (1989) "Residential Land Use Policy and Condominium Housing Development — The Singapore Experience", *Land Use Policy*, April, pp. 121–131.

Sim, L.L., Brown, P., Tan, J.K. and Leong, K.M. (1995) "The Changing Nature of User Demand — A Case Study of Conservation Shophouses in Singapore", Paper presented at the Pacific Asia Property Research Conference, April, Singapore.

Tan, T.K. (1988/1989) *Conservation in Tanjong Pagar: An Economic Perspective*, Unpublished dissertation, National University of Singapore, (Singapore).

Teo, S.E. and Savage V. (1985) "Singapore Landscape: A Historical Overview of Housing Change", *Singapore Journal of Tropical Geography*, Vol. 6, pp. 49–63.

Urban Redevelopment Authority (1988) *Master Plan for Civic and Cultural District (Draft)*, January, Singapore.

Urban Redevelopment Authority (1992) "Planning a Downtown for the 21st Century", *Downtown Core and Portview Development Guide Plans (Draft)*, August, Singapore.

Wong, A.K. and Yeh S.H.K. (eds.) (1985) *Housing a Nation — 25*

Years of Public Housing in Singapore, Maruzen Asia for Housing Development Board, (Singapore).

Wong, M.P. (1990/1991) "Evaluation of Tanjong Pagar Conservation Scheme", Unpublished dissertation, National University of Singapore, (Singapore).

Wong, P. (1993) "Revised Housing Guidelines for Medium and High Density Housing", SIA/SIP Planning Workshop on Strata Bungalows, Cluster Housing and Revised Housing Guidelines, July, Singapore.

Yeung, Y.M. (1973) "National Development Policy and Urban Transformation in Singapore", Research Paper No. 149, Department of Geography, University of Chicago.

2

Towards a Tropical City of Excellence

Liu Thai Ker

Introduction

Singapore is an Asian city in spite of its many modern high-rise buildings. Apart from its geographical location and the ethnicity of its people, it is also Asian in the sense that a large population is concentrated in a small area — 3.02 million people in an area measuring some 640 square kilometres. Island-wide, the population density is 4300 people per square kilometre. Recently, government planners have embarked on its plans to make Singapore a city of excellence.

This paper discusses the vision of excellence for the city. It reviews the current status of urban development and then discusses the efforts required for the achievement of a well-planned and well-developed city. These efforts will need to come not only from government agencies like the Urban Redevelopment Authority (URA) but also the private sector and the society as a whole.

The long-term physical planning of Singapore can be divided into two phases. The first phase started in the early 1960s when government professionals were involved in solving the most urgent and immediate problems faced. After 25 years of nationhood, there emerged a second phase. In this phase, government planners are now more confident and in a better position to make longer term projections. They are also equipped with better resources to achieve these projections. During this second phase, planners are starting to work towards a vision of Singapore as a tropical city of excellence.

In the last 25 years, Singapore has become a city which functions efficiently in support of the needs of the people and the business community. It is a city with the inner vitality and ability to generate ample jobs which will ensure its long-term survival. Singapore is also a city with a friendly and secure community. The creation of a city of excellence, however, means further development and going beyond such achievements of the past.

The concept of a tropical city of excellence is multi-faceted. As part of this vision, one would include greater road access to all parts of the island. There has to be integration of transport and land-use planning with proper interface between public transport facilities and cars. Culture and leisure have to be part of the city plan requiring world-class architectural design. The treatment of urban spaces would be important both in the city and its commercial nodes. Singapore is also a business centre and as such, the planning of the city-state has to ensure that there are adequate spaces for business development with appropriate and up-to-date planning regulations to guide such development. It is the role of planners and agencies like the URA to initiate the urban development process through systematic planning of space and linkages in the city.

This paper will first, provide a quick review of the physical plans for the city of Singapore. It will then discuss five key issues related to the development of the city. These issues concern the effort to decentralise the people and business activities, the desire for a garden city and an island city, greater variety in development and the environment and finally, the goal of a city of excellence.

The Plans

The above plans are the Concept Plan, the Master Plan, the Development Guide Plans as well as the study plans for selected areas, urban design proposals and relevant planning and related regulations. The Master Plan, drawn up in 1955 and approved in 1958, is a legal document which has served as a reference point for the development of Singapore. It indicates the zoning and land-use intensity permissible for any piece of land in Singapore.

To help planners look ahead, the Concept Plan has been introduced. Consultants from the United Nations Development Programme (UNDP) were engaged in the period between 1967 and 1971 to help prepare the Concept Plan for the whole island. The population projected and planned for at that time was four million. Subsequent to the completion of the Concept Plan, the development of public housing, industrial estates and infrastructure has adhered to its specifications, although with modifications and improvements being made in the process. There had been economy of effort because of the close coordination between government agencies, all of which were working to meet government demand. The Concept Plan specified the amount of land needed and the size of the population *vis-á-vis* the development of major roads, expressways as well as public transport facilities such as the Mass Rapid Transit (MRT). With current planning needs, the Concept Plan is being revised.

Greater detail is required to allow both the government and the private sector to better understand the intention behind the development of different parcels of land. The Development Guide Plans (DGP), introduced in recent years, are therefore being drawn up for specific areas. Among the earlier DGPs are those for the city-centred neighbourhood of Kampong Bugis and non-central area community of Simpang. These plans cover some 55 planning communities and they provide detailed information on land-use intensity, height of buildings and setback. After the plans have been fully discussed and amended, they will be given legal status similar to that of the Master Plan. They will then guide both the developers on the types of investment permissible and the architects and engineers in their building design.

In preparing the revised Concept Plan and the DGPs, four broad objectives are addressed. They are to organise the urban structure of the island, review the urban fabric which comprises the road network, the MRT, green belts and water bodies, creatively examine the array of land-use types and also specify ways to upgrade the quality of the physical environment to a level that may be considered excellent. Each of these areas of work will contribute to the creation of a city of excellence.

Urban Structure

In the area of urban structure of the island, the existing five urbanised corridors identified in the Concept Plan will remain. The Concept Plan was updated in 1980 to incorporate the developments in the period between 1972 and 1980 as well as the projections that had been done. In this update, a fifth development corridor towards the Punggol area in the north-eastern part of the main island was added. Originally, there had been four corridors — the East, the West as well as a ring of two corridors which met around Woodlands in the northern part of the island. The fifth corridor had been added in recognition of the population density and the basic infrastructure which were already there. Newly reclaimed land was also included in the Concept Plan. Now, owing to the growing congestion and increasing costs in the Central Area, an effort will be made to encourage decentralisation.

In the latest revised Concept Plan, encompassing the five urban corridors, the island has been divided into five regions — the central region and four more outside of this. Each of these regions will have a population of half to one million people, which would make the region the equivalent of an independent city. Within each region, there will be a regional centre which will be served by expressways, arterial roads and MRT lines.

Having set down the urban structure, there is a need to specify the details. These provide the urban fabric.

Urban Fabric

In re-thinking the urban fabric, there is a need to build upon the solid foundation of the existing urban framework. The road network and MRT routes need to be updated. Significant natural assets such as hills, valleys, green belts, rivers, beaches and the shoreline are being critically re-assessed for conservation. The island will be linked by many different kinds of networks — roads, train lines, green belts and water courses. With the limits on land resources, it is not possible to allocate a lot of land for any one of these features. They can, however, be planned so that they can exist symbiotically among themselves and also with commercial nodes and urbanised belts.

The urban fabric provides the details concerning land uses and transport in the city. Beyond these details, there is a need to set down the parameters for such land uses. Such parameters are of utmost importance in Singapore considering the major constraint faced in physical planning — land scarcity.

Land-use Parameters

For the review of land-use parameters, a careful look is being taken at the amount of land available and projected needs for the year 2030 and beyond. The URA has launched various typological studies. Each land use is being examined in view of the range of different developmental proto-types that are wanted in Singapore. For example, the range of commercial developments would include downtown, regional, town and neighbourhood centres as well as coastal resorts and even village centres.

Planning parameters are ascribed to every type of development to make each recognisably different from the others. The typological studies have been repeated for industries, facilities, amenities and open spaces. There is also a need to create new types of land uses so that Singapore can keep abreast of developments elsewhere in the world. There are studies being done on new housing types, mega-malls, business parks, theme parks and new forms of leisure and resort developments.

After these typologies have been created, they have to be test-mixed to see how much of each type Singapore can afford to have. Since the land supply is limited, care is required in planning in order to give every land-use typology a place, albeit in varying proportions. This will, in turn, create a greater visual variety and choice of environment. Such variety and choice will arrest the tendency in a city which is short of land towards denser development intensities and hence, similar environmental characteristics over the entire island.

Quality of Excellence

The fourth area covered by the plans is the quality of excellence. This is achievable by highlighting and enhancing the magic of the tropics — sun, flora and fauna, hills and rivers, beaches and sea, old monuments and traditional architecture, civic spaces, boulevards as well as covered and uncovered pedestrian links. The achievement of excellence must be seen in relation to the aspirations of Singaporeans. A better range of choices must be offered.

The city has to satisfy the cultural and spiritual needs of the people and combine this with the sense of security and well-being. Singaporeans are no longer only contented with a clean, green, well-functioning and economically thriving environment. Young people are talking about identity, historical roots, self-expression and demanding opportunities to contribute to developments in literature, music, dance, art, crafts and architecture. A whole array of new business districts and cultural districts would have to be provided.

In the planning process, the emphasis has been on ensuring the availability of land for various developments. With the limits on the supply of land and population growth, the only variable that permitted flexibility was the density of development. Moreover, beginning with the 1960s through to the 1970s, the preoccupation was with the rapid construction of much needed housing, commercial and industrial floor space as well as the necessary infrastructure. In the 1970s, emphasis started to shift to facilities and amenities while in the 1980s, more attention was paid to institutional and cultural buildings.

Decentralisation of the Population

Singapore is a modern city that has been built from what had been a very backward, unhygienic and extremely congested place. Originally, its 1.6 million people were all concentrated in dilapidated dwellings located within two to three square kilometres of the Central Area. From the beginning, it was obvious that there was a need to decentralise the population. Under the Singapore Improvement Trust (SIT), the colonial government only developed 27 000 dwelling units over a 32-year period.

History of Resettlement and Public Housing

In 1960, the new Singapore government assessed that in the period between 1960 and 1970, 110 000 units of flats would be needed. Since then and until the mid-1980s, the rate of building has continually been stepped up. By 1990, the total number of public housing units was 615 000. Public housing has played a key role in decentralisation.

The initial emphasis of the public housing effort had to be on quantity in the delivery of alternative accommodation. Conditions in the slums and squatter areas were then so bad that anything else which provided sunlight, cross-ventilation, electricity, water and sewerage meant a marked improvement in the living conditions of a majority of the people. Basically well-planned but utilitarian type of construction helped spread benefits to the maximum number of people in the shortest possible time.

Over the last 25 years, more than 200 000 squatters and farmers have been resettled to public housing and industrial estates built by the Housing and Development Board (HDB). Through such resettlement, it has been possible to clear land which was strategically located and highly suitable for development with higher economic value.

To make resettlement as palatable as possible, people were accommodated as near to their original residential addresses as possible. Calculations of affordability for housing as well as business premises had to be realistic. Industrial buildings of the early days were then built according to basic construction standards so the businesses could afford to move into them. Generally, the people who were resettled had been living and operating in places where there had been negligible or no rent. In their new accommodation, they not only had to pay rent or housing instalments but also bills for water and electricity.

Not every business that has been resettled has succeeded although failures have been kept to a minimum. The proprietors of some businesses which would have had a marginal existence to survive in squatter areas, simply do not have the business acumen to operate in a new environment. So, the way in which the government has structured resettlement compensation was important if not crucial to the success of the population decentralisation effort. Cash compensation was introduced so that business proprietors could take this and change vocations.

During the earlier phase of planning, conscious effort was made so public housing development would be mostly in the form of self-contained new towns to complement development elsewhere on the island. Emphasis was on facilities, convenience and accessibility. Residents have access to employment in the industrial estates located in and near new towns. The development of new towns contributed to the improvement of the urban environment as well because of the more even distribution of green spaces, shops, schools and such amenities. There is a hierarchy of green spaces in new towns — town parks or gardens and neighbourhood parks.

Public Housing and National Development

Now, in the effort to bring greater variety to the environment, private developments have been planned in the HDB new towns. The intention is also to reduce the social and visual differentiation between private and public housing.

The public housing programme has been an integral part of the overall development of the country. It has transferred Singapore into a property-owning society. By 1990, 87 per cent of the population in Singapore was living in public housing. Home ownership has been extended within two decades to more than three quarters of public housing residents. Homeownership increases citizens' stake and enhances their sense of being rooted in the country. With most of the housing estates built on land populated previously by squatters, the public housing programme has resulted in a massive redistribution of people and the residential integration of different ethnic and socio-economic groups.

Decentralisation of Business Activities

A second big wave of decentralisation can be expected in the near future. The city centre has been getting so big, property values have escalated and congestion again is becoming a problem. Serious thought is being given to decentralisation of some businesses to areas outside the city centre, that is, to regional, town and neighbourhood centres as well as other commercial nodes. In this way, alternatives in business locations are provided.

The plan is to encourage back-room type of businesses to move to the regional centres. Service- and market-oriented businesses can stay in the central area. Businesses which are research-oriented and hence, require neither the support facilities of the city centre nor those of regional centres can be located in business parks in a garden-like environment, many not far from the MRT stations. Effort will be made to encourage the use of public transportation in order to reduce road traffic. This does not mean that no more roads will be built. There will be more roads but the intention is also to explore and exploit to its fullest, the potential of the MRT and buses, that is, public transport.

Island City: The Garden-City Theme Revisited

Anti-pollution measures instituted have been highly successful. Few cities have achieved an urban environment as clean as Singapore's (Ooi 1995). Without incurring the use of too much land, ways have to be found to make the city as much of a Garden City as possible. There should be greenery and more water everywhere. Special attention will continue to be paid to parks and the conservation of green areas at critical points. Both the Concept Plan and the Development Guide Plans incorporate the injection of green areas where they are lacking. Planners are working towards a vision of a green network island-wide.

Now planners are considering the second cycle of urban improvement which will be aimed at quality, variety, excellence in architecture and the design of urban spaces. Better use should also be made of natural assets such as the tropical weather and lush vegetation. Future plans for the Botanic Gardens, for example, call for the conservation of the existing gardens, extensions with a new garden area of a different design, and upgrading of its scientific status and scholastic value. In the development of Fort Canning Park, there will be an attempt to relate it to the early history of Singapore. Effort is also being made to preserve the natural habitats of plants and wildlife on the island.

Since Singapore is surrounded by water, more thought should be given to ways of making water part of urban development. Marina South, which is to be the Central Business District of the 21st century, the Singapore River, Kampong Bugis and

Tanjung Rhu are all intended to exploit the advantages of their waterfront locations.

In the bid to make Singapore more of an Island City, there can be plans to bring housing closer to the water and the development of commercial and other leisure activities at waterfront locations — the sea, river and even canals. The major rivers and canals are now reasonably landscaped and are safe for water sports. New beaches near the waterfront downtown provide attractive venues for boating and water-skiing. A combination of vision and planning skills can bring the city closer to being a vast tropical resort. Life in the city streets need not be incompatible with the serious business activities taking place inside the city's buildings.

Variety in Development

The decentralisation of business activities and the strategies being used to create a Garden City and an Island City will lend greater variety to development in different parts of the island. Conservation of natural habitats as well as historical buildings helps us retain an important dimension in the city environment. Another effective way to enrich the personality of the city is through a conscious effort to allow pockets of existing development to retain their character and nature.

Singapore is a newcomer in preservation and conservation. It is, nonetheless, the first newly industrialising country to have a comprehensive Conservation Master Plan which is being implemented. Four main groups of buildings have been identified for conservation or preservation — institutional civic buildings which are mainly government buildings located in the Civic District, religious buildings, bungalows and shophouses. The treatment of each of these groups of buildings differ. So does the process of preservation and conservation.

A number of great cities have world-class historical and architectural assets. The architectural heritage of Singapore cannot match them in significance. Therefore, we will need to compensate for this with excellent new architecture, squares, plazas and gardens. The aspiration of making the city more pedestrian-friendly will also help provide the variety sought in city design. This is expressed through planning, an example of which is the Civic District Plan. Part of the district has recently been redeveloped to incorporate historical building sites and landscaped open spaces as well as streets and pedestrian footpaths.

A very close look will be taken at the variety of developments through typological studies of land use. A wide range of different types of environment within each land-use type will be created and identified as the basic ingredients of the Concept Plan. While Singapore cannot possibly match some of the great cities which have many centuries' worth of headstart, an impact can still be made with the creation of good public spaces, beautiful gardens and cultural and institutional buildings.

The upgrading of existing educational buildings can, potentially, play an important role in giving architectural prestige and introducing gracious living amidst the activities of the city. In July 1991, the Nanyang Technological Institute became Nanyang Technological University with a new look for its campus. Upgrading will be also be carried out on the National University of Singapore campus. The Temasek Polytechnic campus is being developed at Tampines. Better facilities and design are also slated for privatised schools. A museum precinct has been designated. There will be a Singapore Performing Arts Centre. In many of these developments, international architects and the best of local architects will be consulted.

To cap it all, Singapore's multi-ethnic society makes it a colourful city. In terms of planning, Chinatown, Arab Street area, Little India and Geylang Serai will retain their respective ethnic flavours for a long time to come. In this particular aspect, some overseas visitors have urged urban planners to make Singapore a garden of different cultures, much like a natural garden of flowers. Given proper nurture, every pocket in the city can blossom and contribute to a great variety of colours, shapes, sizes and fragrances.

A Tropical City of Excellence

A city could have everything right and yet still fail to qualify as an excellent city. In the final analysis, it is judged by its appearance, architecture, urban design and public spaces. Planners and architects have, therefore, to work together to find ways for new buildings to fit into and relate to the overall urban fabric, to be beautiful buildings in themselves as well as reflect our multi-ethnic culture and Asian heritage. The buildings should also reflect the Singaporean lifestyle, climate and local conditions.

Excellence has to be built upon two elements which are basic to a city's existence — survival and refinement. Singapore is an island state with a relatively large population size. As a result, a lot of it is urbanised with many high-rise structures.

The demands on land are numerous. The social needs of the people have to be accommodated — adequate housing, places of work and schools. Political consideration such as an inculcation of a sense of ownership is critical. The city has to facilitate economic growth. The Central Area has been revitalised and will gradually become a cosmopolitan downtown. Industrial estates have been built to attract local and foreign investors in order to ensure full employment. Infrastructure has also been rapidly developed and environmental pollution control measures strictly implemented so that the city can function well and be relatively free of health hazards. Survival issues have been taken well in hand. They are a fact of life in Singapore and will continue to be part of planning considerations.

The city, however, cannot remain at the survival level and become excellent. The process of upgrading has started so Singaporeans can enjoy a more gracious lifestyle. Public housing estate planning and building design are already being upgraded. Institutional amenities and social and recreational facilities are increasing in number and improving in quality to meet higher aspirations. Comfort and convenience will be elevated through better quality and intelligent buildings as well as through the nurturing of a more pleasant environment and an integrated transportation system.

Cities are considered great for different reasons. The old cities in Europe, Africa and Asia like Venice, Cairo, Rome or Istanbul are great because of their rich historical, cultural and architectural heritage. More modern metropolises such as Chicago and New York are great because of the new urban images they have created since the late 19th century. Yet, not all these cities have been able to maintain a high environmental quality.

Singapore cannot claim the rich heritage of the historic cities nor the glamour of modern metropolises. However, it can attempt to become an excellent city of the 21st century in a relatively short period of time. Singapore should maintain its high environmental quality. It is a city which is able to promote a gracious lifestyle. In addition, it has its unique though modest historical heritage. There is a rich endowment of natural flora and fauna. Singapore has a colourful multi-ethnic culture. It is one of the first cities in the tropics to attempt to achieve a modern architecture which is both Asian and tropical. By constantly trying to better what has been achieved, a unique city of excellence can be created.

REFERENCES

Government of Singapore (1991) *The Next Lap* (Singapore: Times Editions).

Liu, T.K. (1990) "Improving the Living Environment of Singapore", Unpublished paper presented at the Fourth Symposium on Our Environment, 21–23 May, Singapore.

Ooi, G.L. (e.d.) (1995) *Environment and the City — Sharing Singapore's Experience and Future Challenges* (Singapore: Times Academic Press and the Institute of Policy Studies).

Urban Redevelopment Authority (1991) *Living the Next Lap — Towards a Tropical City of Excellence* (Singapore).

3

Town Councils in Public Housing Estates: Change and Implications

Ooi Giok Ling

Introduction

The success which has been the story of public housing in Singapore has been well-documented not only by the Housing and Development Board (HDB) which is the government statutory board that plans and provides public housing but also researchers who have evaluated the programme (Phang 1992; Pugh 1989; Field 1987). While the most tangible aspect of the success lies in the number of public housing apartment units which has been developed by the HDB since 1960, the management of these public housing estates has had an equally visible impact on the built environment in Singapore.

The decision-making in both planning and development of public housing has been highly centralised. Similarly, decision-making powers on public housing including its management until recently, have been the HDB's. Centralisation of planning policy decision-making has had a number of implications for the built environment. This is especially the case with the planning and development of public housing estates in Singapore as the following section intends to illustrate.

It focuses on the impact of centralised decision-making on the planning and provision of public housing estates. This discussion will highlight the impact of planning standards and practices adopted by the HDB on the built environment in public housing estates.

The section on the management of public housing estates will provide the context for the next section which will focus on the introduction of Town Councils. Town Councils were introduced with the aim of localising the management of public housing estates. In this section, its aim is to understand the impact of the changes involved with the shift from the centralised management of the estates by the HDB to Town Councils. The implications arising from the changeover for public housing estates and its residents will also be discussed.

The Planning and Development of Public Housing Estates since 1960

Since the decision was made to help relieve the congestion in the central area of the city through the decentralisation of most of the people living there, much of the residential estate development has been concentrated outside the city centre (see Chapter 2). The congestion in the central area was a result of both rapid population growth as well the thriving of the former British colony as an entrepot port.

Congestion was also due in part to the largely unplanned nature of the city's physical development. According to the *Master Plan: Reports of Study Groups and Working Parties* (1955, p. 1), "Land has been developed to meet requirements as they arose ...". Not surprisingly, the result by the later 1950s was an outdated street pattern affording only poor access to the city centre, overcrowded and dilapidated housing and an inadequate provision of urban facilities and services including schools, open space, clinics, hospitals and other amenities.

Urban redevelopment was urgently needed but while policy-makers recognised that the problem had to be addressed on an urgent basis, such redevelopment was not able to proceed without the freeing of land in the city centre. The public policy decisions that led to the relocation of people and major land-uses from the central area will be discussed in some detail in Chapter 5.

Briefly, redevelopment of the central area was only able to proceed with legislation to make land available. Such legislation also empowered the state agencies to acquire land for urban redevelopment. Much of this land had previously been divided into very small lots for many private owners. The land acquisition powers of the state agencies circumvented the stalling of the redevelopment programme by the necessity of obtaining the cooperation of the owners of adjacent land plots.

Apart from urban redevelopment, the study groups for the Master Plan had also recommended the development of the other parts of the island colony. These developments, according to the recommendation, should provide low-rental accommodation, local employment, community and entertainment facilities in order to attract people from the central area.

The public housing programme which was launched in 1960 with the establishment of the HDB, appears to have followed the recommendations rather closely. In planning and developing public housing estates, however, the HDB has been adjusting and revising standards and practices to accommodate changing demands over time (Ooi 1991). Such changes in demand for public housing over the last three decades have been evident not only in the eligibility criteria to qualify for public housing but also the types of housing units developed and the planning standards and provisions implemented.

Income ceilings on the eligibility for public housing have been revised upwards on several occasions in the last three decades. This is not surprising, considering the growth in income levels between 1960 and 1990. While the per capita GNP was US$800 in 1965, it had increased to US$12 543 by 1990. The growth in incomes has also been paralleled by the increasing demand for larger areas of living space. The HDB's response has been to develop larger flats over time as illustrated in Table 3.1. In the table, the ages of the new towns that have been developed are shown together with the composition of apartment types that have been provided in each town. Clearly, there has been progressively a shift from the smaller three-room flats to the larger four-room and five-room flats.

The HDB's quinquennial building programmes are listed in Table 3.2. These programmes give a perspective of the scale of the public housing effort in Singapore. Since 1960, the HDB has developed more than 600 000 units of public housing apartments. Some 86 per cent of the people in Singapore now live in public housing estates.

Table 3.1 Types of Public Housing Apartment Units Developed by the HDB

New Town	Year of Construction	Types of Apartments 1-room	2-room	3-room	4-room & larger
		%	%	%	%
Queenstown	1952	7.0	21.3	68.9	2.9
Toa Payoh	1965	31.3	15.6	42.5	10.6
Ang Mo Kio	1973	5.5	11.0	54.0	29.5
Tampines	1980	0.5	0.4	39.4	59.7
Bishan	1984	—	—	16.7	83.3

Source: Housing Development Board (1984/85, 1986/87, 1988/89) *Annual Reports;* Tan et al. (1985), p. 99.

Table 3.2 HDB's Building Programmes

Years	Total Number of Dwelling Units Built
1960–1965	53 777
1966–1970	63 448
1971–1975	110 362
1976–1980	130 981
1981–1985#	189 299
1986–1991	130 160

includes units built by Urban Redevelopment Authority.

Source: Housing and Development Board (1991/92) *Annual Report*, p. 82.

In terms of the delivery of public housing to meet people's basic needs for accommodation, centralisation of decision-making has clearly been an advantage. Being the sole public housing authority, the HDB has been able to respond swiftly to changing needs and conditions. There has been a cost as well. This becomes evident in the following section.

Since the land area of Singapore measures only some 620 square kilometres and it has a population of 3.02 million, the solution to the provision of adequate housing has had to contend with the competition for land resources. To accommodate both the demand for housing and changes in such demand over the last three decades, public housing estates have been high-rise and high-density developments. The high-rise and high-density formula has been an advantage from the planning point of view as the following section intends to illustrate.

The Planning Standards and Practices in Public Housing Estates

Centralisation of decision-making on the planning and development of public housing estates has meant a uniformity of planning standards and practices. Not only are the estates more or less similar in appearance, provision of facilities and services has been virtually identical in all the estates.

Basically, the planning of public housing estates has been operating within a physical framework that has been delineated largely by population "catchments" for urban facilities and services, comprising mostly retail shops, fresh food markets and schools among others.

The planning standards adopted by the HDB for its estates and new towns have been based on a hierarchy of planning units and a spatial distribution of facilities and services that has emphasised equatability among the estates and new towns. This hierarchy comprises new town, neighbourhood and local planning units. Each of the units would be planned around a service centre, the size and functions of which would be determined by the number of dwelling units and hence, families being served by this centre.

Differences have emerged largely as a result of the revisions in planning standards and practices that have occurred from time to time. Such revisions have been necessitated by the planning of new towns that had to be relatively self-contained because they were located at increasing distances from the central area. Earlier estates that had largely been located in and near the urban fringe could depend on the central area for services. Changes, however, have also been encouraged by the success that has been achieved in meeting basic housing needs and the growing necessity currently to respond to demands created because of the residents' rising affluence.

The following review of the range of facilities that have been added to physical provisions in public housing estates reflect the changes that have been made:

> To make the new towns of 150 000 to 300 000 people self-sufficient and to keep up with the residents' changing needs, the HDB has had to continuously increase the range of facilities. ... Among the commercial premises, cinemas which first appeared in the HDB estates in the late 1960s were supplemented with emporia and supermarkets in the 1980s. ... In addition to swimming complexes, sports complexes and indoor stadiums, steadily appear in the estates. Parks and gardens as well as jogging tracks, keep-fit corners and children's playgrounds find their way into every corner of the housing estates. To meet high parking demands, garages and specially designated bus and lorry parking lots are provided in addition to open parking lots. And free ground floor spaces of the residential blocks are converted into education centres, child-care centres, senior citizens' clubs, community sub-centres, neighbourhood police posts, offices for the residents' committees and community halls (Liu 1985, p. 13).

While changes have been seen in the planning standards and practices, there has been equal emphasis on the standardisation of planning and design of the public housing estates. "A great deal of effort has been given to standardising the land-use distribution, the density and the components of all the housing estates. Rigid specifications for the layout plan of a flat or a building block are also imposed. Standardisation is ... a necessary compromise if the quality of housing is to be kept consistently high ... the buildings are designed for an industrialised method of construction at least in concept, with all the consideration for simplicity, repetition and mass production" (Liu 1975, pp. 125–126).

The spatial distribution of the facilities and services provided in public housing estates is illustrated in Table 3.3. This table also shows the hierarchy involved in the distribution of the facilities and services. Every new town, therefore, has a hierarchy of precincts, neighbourhoods and neighbourhood centres as well as a town centre. Each precinct has its focal point and range of facilities allocated to this spatial unit. There is considerable regularity in the distribution of facilities in neighbourhoods as well, including the neighbourhood centres. Similarly, every new town is served by a town centre as well as a range of urban facilities and services such as a bus terminal, town garden or library.

Given the emphasis on planning for equatability and standardisation, it is not surprising that some new towns and estates are difficult to distinguish by appearance. Hence, the observation that "... towards the end of the 1970s — six new towns started construction in quick succession. It is clear that with their simultaneous construction, the new towns cannot but acquire almost identical appearances" (Tan et al. 1985, p. 98).

In spite of the emphasis on standardisation, changes in the planning standards which have guided the development of physical provisions for new towns and public housing estates have led to the emergence of differences among these areas. These differences would exist among new towns developed earlier and those which were begun more recently.

An upgrading programme has been introduced to bridge the differences between estates developed earlier and those which are more recently developed. Through this programme, the older estates have been clustered into new town units and an up-to-date range of estate facilities and services added.

The upgrading effort, however, has been to enable the HDB to address the differences in physical provisions that have developed between the older and newer estates and new towns. Such upgrading effort has extended to the demolition of blocks of smaller apartment units and their replacement with blocks of larger units that are more in demand among young home buyers.

Table 3.3 Distribution of Public Housing Estate Facilities and Services

Retail Facilities	**Floor Area (m^2)**	**Planning Standards (Shop to Dwelling Units)**
Shops	30 – 400	1 to 70
Kiosks	5 – 15	1 to 600
Emporia	4500 – 6500	1–2 per New Town
Supermarkets	1200	1–2 per New Town
Eating Houses	450	1 to 750
Restaurants	90 – 2000	1 to 1000
Mini-Markets	450	1 to 6000
Wet Market-cum-Cooked Food Centre*	—	1 to 5000
Market Produce Shops	130	1 to 6000
Market Produce Lock-up Shops	40	1 to 500
<u>Other Commercial Facilities</u>		
Cinemas	1800	2 per New Town

Institutional Facilities	**Area (ha)**	**Planning Standards (Facility to Dwelling Units)**
Primary Schools	1.8	1 to 2300
Secondary Schools	2.7	1 to 4100
Junior Colleges	6	1 per New Town
Vocational Institutes	6	1 per New Town
Libraries	0.3 – 0.4	1 per New Town
Polyclinics	0.5	1 to 30 000
Community Centres	0.4	1 to 4000 – 5000
Mosques	3000 m^2	1 per New Town
Churches	0.3 – 0.4	1 to 12 000
Hindu Temples	0.2	1 per 2 New Towns
Chinese Temples	0.2	1 per New Town
Residents' Committee Centres, Community Halls, Neighbourhood Police Posts, Kindergartens, Child-Care Centres, Senior Citizens' Clubs.	0.2 – 0.4	6 – 7 reserve sites for these facilities per neighbourhood

Source: Tan et al. (1985), p. 103.

The emphasis on standardisation and uniformity notwithstanding, planners and designers of public housing estates have recognised the importance of identity and differentiation in new town planning and design. "In recent years, visual identity has become a vital component in the HDB planning and design process" although "a guiding principle has been that variations must always be strictly controlled. This is necessary not only to keep costs under control but also to avoid the visual chaos that would result if each component of a new town is given its own visual identity in isolation, without regard to the overall image" (Tan et al. 1985, pp. 106–107).

In short, there has been recently, an attempt to allow for greater variation and variety among the new towns. Yet, as the above discussion has emphasised, such variation and variety has been carefully controlled by the HDB. This is aptly reflected in the observation that "... the repetitiveness inherent in much of the HDB's work is used to constitute a backdrop, against which variations, or foreground elements, are then selectively introduced to good effect" (Tan et al. 1985, p. 107).

In adjusting its planning standards and practices to accommodate changes in demand and the resources available, the HDB has apparently also relied on feedback and research findings (Liu 1975). Obtaining such feedback has been facilitated by the fact that the HDB also managed the public housing estates that it had developed.

The Management of Public Housing Estates

The management of public housing estates by the HDB has involved essentially, two major components. One component is the allocation of the housing and commercial units developed by the HDB. This means the HDB also set the eligibility criteria that would have qualified the residents and businesses for both public housing as well as commercial premises in public housing estates.

For the majority of the people and most of the businesses being relocated from the central area, the HDB would have provided alternative accommodation in public housing estates. Resettlement in public housing estates developed by the HDB would have formed part of the compensation package offered to people and businesses being relocated from the central area.

The other component of the estate management functions provided by the HDB involved the cleaning and maintenance work in public housing estates. This work would comprise refuse removal and the daily sweeping and washing of common areas and facilities in public housing estates as well as repairs and maintenance of essential services — lifts, water and electricity supplies. Through the years, the HDB has developed a round-the-clock emergency repairs and maintenance service. Estate management work also included the provision of landscaping and maintenance of such landscaping work in the public housing estates.

Apart from the economies of scale realised through the centralised management of public housing estates, centralisation of decision-making was advantageous because the authority for both policy formulation and implementation was effectively vested with one organisation. Technically speaking, this should make the bureaucracy more responsive to changing needs and conditions.

The HDB had managed its estates through a spatially decentralised system of area offices (see Fig. 3.1). Each area office had under its charge some 15 000 public housing apartment units and a number of other commercial premises. These area offices were each located in a specific part of a new town or estate.

Exchange between area offices and the HDB headquarters was institutionalised as were the links between the public housing residents and the area offices. Table 3.4 shows the list of the range of services and activities that had been organised by the HDB to link it to the residents. Both the area offices and the services listed in the table had served to make the estate management functions of the HDB more accessible to the public housing residents. These area offices made the running of public housing estates more decentralised than otherwise perceived.

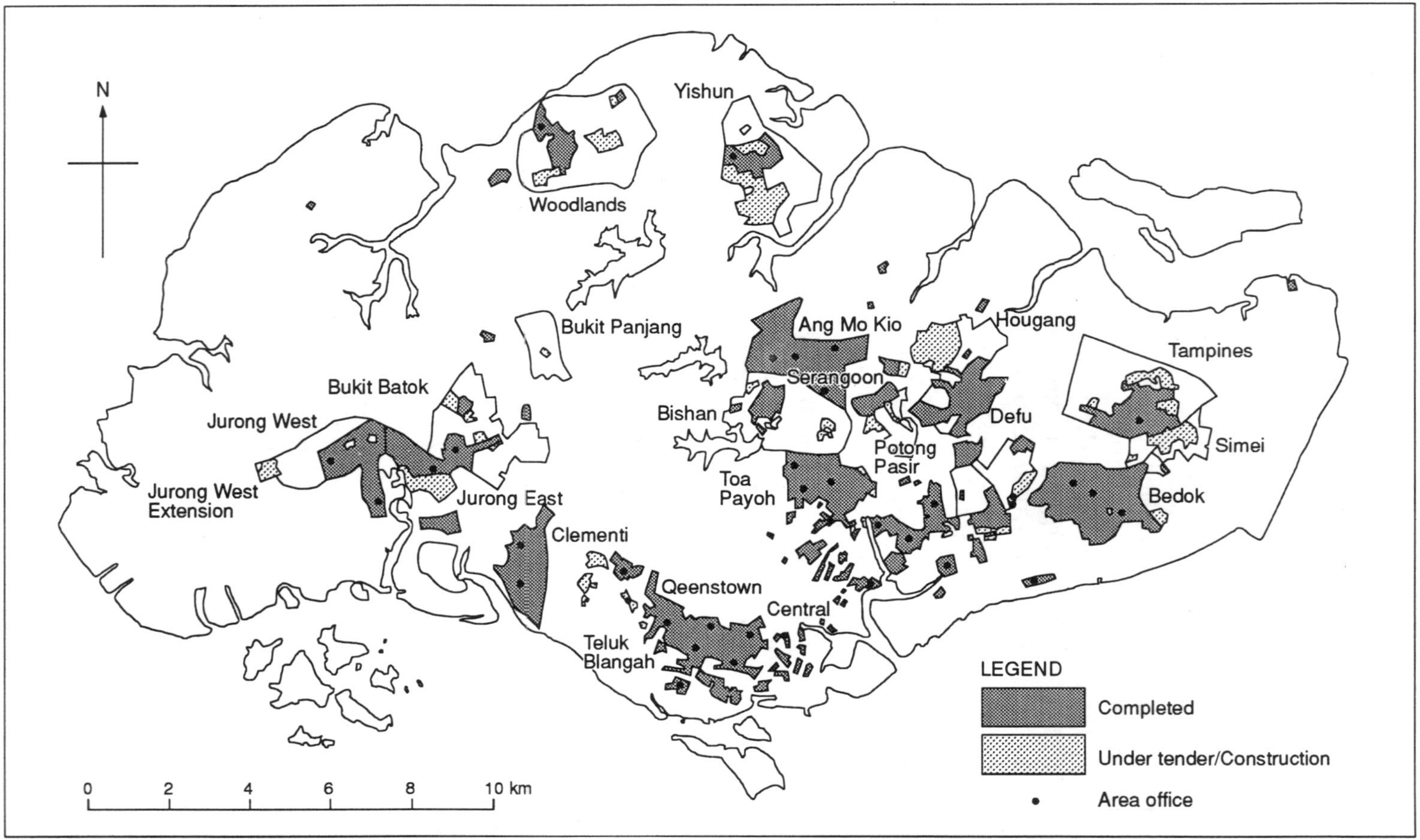

Source: Redrawn from Wong and Yeh (eds.) (1985) *Housing a Nation – 25 Years of Public Housing in Singapore.*

Table 3.4 Residents' Programmes and Activities Organised by the HDB

Type	Item	Frequency	Workload in 1984
1. Education & communication	*Our Home*	2-monthly	433 400 copies per issue
	Press releases	Weekly	—
	Area Office Day	Yearly	—
	Residents' Handbook	To every HDB family	—
	Audio-visual show	– do –	—
2. Liaison with residents	Visit to new residents within 3 months after allocation	On-going	—
	Scheduled visits to flats		180 000 visits
	Participation in national campaigns		300 campaigns[1]
3. Residents' welfare services	Neighbour disputes	On-going	1 080 cases
	Referral services to social agencies		192 cases
	Management of hardship cases		78 cases
	Void deck space allocation for communal use		72 cases
4. Joint campaigns with grassroots organisations	Anti-killer litter	On-going	246 campaigns[1]
	Bulk-refuse removal		22 campaigns[2]
	Clean up		20 campaigns[3]
	Use of plastic bags		12 campaigns[3]
5. Liaison with grassroots organisations		On-going	—

Note: [1] Total number carried out by 36 area offices.
[2] In some area offices, this campaign is held in conjunction with the anti-killer litter campaign.
[3] Campaigns started in late 1984.

Source: Liu (1985), p. 25.

The centralisation of decision-making had a cost because similar rules and regulatory procedures would have been applied across all the estates regardless of their differences. Decision-making was also the prerogative of the bureaucrats and the public housing residents had relatively little say. Residents' Committees (RCs) were formed, comprising members from the public housing residents. When it was first established, the RC Secretariat was, however, located in the Ministry of Community Development and officers from the HDB were seconded to the Secretariat to serve as RC Liaison Officers. The RCs appeared to have served more the role of explaining the HDB's policies rather than a mechanism for soliciting residents' views on public housing policy issues.

The decision to introduce Town Councils to manage public housing estates suggests that the government's objective was the localisation of estate management beyond what had been achieved through the spatial decentralisation of services by the HDB. In the following section, the discussion will focus on the changes involved as well as the implications for public housing and its residents.

The Introduction of Town Councils

The Town Councils Act was introduced in 1988. Prior to the passing of the Act, there had been a pilot project located in a new town. This pilot project established three Town Councils in the new town which took over the management of public housing estates from the HDB. Town Councils are "... designed to give Singaporeans the chance to decide for themselves the kind of environment they prefer and to let them participate in the day-to-day running of their estates" (Ministry of National Development 1988, p. 1).

Town Councils are each chaired by a Member of Parliament (MP) because the councils manage public housing estates in every political constituency. The boundaries of the "towns" managed by each Town Council would, therefore, be defined by those of electoral boundaries.

Town Councillors would be people nominated by the MP chairing the council. There is a requirement that at least two thirds of the councillors would have to be residents from the public housing estates which the council is managing. These councillors hold voluntary positions although their posts are considered public office.

The management functions of the Town Councils are much narrower than those carried out by the HDB. Town Councils are required to keep the estates clean and in good repair. Furthermore, they need only manage the public housing apartment units and common areas as well as facilities.

The HDB continues to be the lessor and land owner in public housing estates. This means that the statutory board maintains its functions of leasing and renting properties in public housing estates. It continues to provide such services through branch offices. A number of the former area offices have been rented to Town Councils.

To carry out their estate management functions, Town Councils collect service and conservancy fees from the residents. The Councils also receive grants-in-aid from the government. Such grants have been graduated to reflect the sizes of the apartment units the councils are managing as well as the ages of these units. The principle would be that the smaller units with lower-income households as well as the older flats require more financial assistance with repairs and maintenance costs. Town Councils, therefore, rely on the government for part of their funds. Apart from the charges that they collect from the residents to clean and maintain the estates, Town Councils have few other sources of generating revenue.

The powers that the Town Councils have been given to make and enforce laws are also relatively narrow and are in keeping with the narrowness of the functions that they have been assigned to carry out. Town Councils can make by-laws but only in such estate management areas like those involving the collection of service and conservancy charges or routine management matters such as the keeping of pets by the residents.

On the other hand, Town Councils are required to obtain the permission of the HDB for development plans including the addition of facilities. It appears that the changes the councils can make to the "towns" they manage are relatively limited, even when they possess the inclination to do so.

Participation by residents in the running of their estates has been as carefully defined as the range of management functions being assumed by the Town Councils. Residents participate as councillors, members of committees or sub-committees formed by the councils as well as through their choice of MP since this MP would be in charge of the management of their estates and hence, the value of their homes and finances for their estates. Participation in the decision-making in this manner, however, involves only the public housing residents nominated by the MP who is chairing the council. There are few other forms of participation which are open to the residents. Members of the RCs are invited to Town Council meetings but like the councillors, their views do not necessarily reflect those of the residents as a whole.

According to the Ministry of National Development, "It would be in the interest of residents to be very careful whom they choose to be their representative in Parliament. ... By voting for honourable, capable and dedicated MPs, and by becoming involved in their community, residents will be able to avoid having to pay the price of a badly-run town council" (Ministry of National Development 1988, pp. 1 and 15).

The advent of Town Councils has taken place at the time when after some two decades of one-party rule by the People's Action Party (PAP), opposition party MPs were first voted to office. There are now two opposition party MPs. They were voted to office after the introduction of Town Councils in a trend that seemed unrelated to the performance of the councils in estate management, as envisaged by the government. Voter accountability is, however, being emphasised with the introduction of Town Councils. Emphasis on such accountability would seem to be equal to that of encouraging residents to participate in the day-to-day running of their estates. Since some 86 per cent of the people in Singapore are public housing residents, such voter accountability could be expected to have a significant impact on voting behaviour.

Implications

Town Councils hold out the prospect of self-determination for the public housing estates in Singapore. For the time being, however, the provision for an exercise of choice and preference by the residents faces a number of constraints. Cost considerations would top these constraints. There are others such as technical and professional expertise or at least full-time expertise for the implementation of changes, mechanisms and resources for establishing the preferences of residents and the impermanence of Town Council boundaries.

Cost considerations will be important to Town Councils in implementing changes. The government has emphasised that it will not bale out councils which run into financial problems.

Since each council manages a total of public housing apartment units which would be far smaller than those managed by the HDB, most of the councils will not be able to realise the kind of economies of scale that the HDB was doing. The sizes of the "towns" managed by Town Councils range from as small as 5 000 public housing apartment units to more than 30 000 units. Furthermore, if the objective of introducing Town Councils is to allow the residents to exercise choice and preference in making their estates different from others, then it is unlikely for economies of scale and benefits from mass production to be realised from the future decisions reached by Town Councils for their estates.

Most of the Town Councils have contracted the management work to agents. A few have hired their own estate management teams including the Town Councils run by opposition party MPs.

After Town Councils were introduced, the HDB formed two subsidiary estate management firms which contracted their work from the Town Councils. The HDB's estate management staff were given the option of joining these subsidiary companies. Effectively, the estate management firms operate at a level that puts them between Town Councils and the residents and hence, separating Town Councils from the residents. This has, justifiably, led to some concern that the councils could potentially develop into mini-HDBs and merely add another tier to the bureaucracy already governing the development and management of public housing.

Town Councils and Residents' Participation

Few mechanisms have been added by Town Councils to those already in place in order to keep in touch with the views and preferences of the residents. Apart from the councillors nominated by the MPs and members of committees and sub-committees convened to advise the councillors and Town Councils, council meetings also include representatives from grassroots organisations such as the RCs. These committees first started in 1978. Only residents qualify for memberships and public housing estates have been divided into zones of some seven to eight apartment blocks. Each zone is represented by a RC.

RCs and other such grassroots organisations have made the effort to ensure a multi-ethnic and socially-mixed representation in its membership. According to the former Minister for National Development, Mr S. Dhanabalan, "Community facilities were built in every housing estate to meet the diverse needs of the ethnic groups who were resettled from their traditional areas. Community organisations like the CCCs (Citizens' Consultative Committees), CCMCs (Community Centre/Club Management Committees) and later the RCs (Residents' Committees) brought together local leaders of different racial and social backgrounds. These leaders were able to see the needs of the community as a whole and at the same time, appreciate the specific concerns of each racial and social group in the community" (*Parliamentary Debates* 1989, cols. 650–656).

Grassroots leadership, nurtured by organisations such as the RCs, had since the mid-1980s, increasingly expressed the wish for more autonomy in making decisions regarding the running of their estates (Ooi, Siddique and Soh 1994, p. 47). It was such views which culminated in the introduction of Town Councils because they had the support of the Prime Minister, Mr Goh Chok Tong, "The HDB makes housing rules for two million people. Such over-centralisation is undesirable. While the HDB has discharged its responsibilities admirably, it has done this at a price. This price is uniformity and a rather inflexible set of rules" (*Straits Times*, 29 June 1988).

In part, because of the manner in which residents' participation has been defined, the general public including the residents interested in participating in town council work, is likely to perceive that such participation is only possible through nomination by the MP or subject to membership in the Residents' Committees. Since Town Councils are run by MPs from both the ruling party and other opposition parties and councillors are nominated by the MPs, this might have given the impression that council participation is partisan and residents would have to join the political parties in order to participate.

The Town Councils have conducted surveys to establish preferences among the residents living in the estates they manage but these are expensive to repeat on a regular basis. Newsletters which most Town Councils circulate to the residents tend to do what HDB circulars have done in the past, inform the people about decisions which have been made by the council members. Such practices, since they depart little from what has been happening in the past and are the kind of centralised practices with which the residents have become all too familiar, might have the negative impact of discouraging wider participation by the residents in the management of their estates.

Town Council boundaries can be changed during general elections. This happened to the boundaries of some "towns" during the 1991 general elections. In part, the reduction in the number of councils from 27 to 24 after the last general election reflected the reality of operating with economies of scale and the need for larger "towns" with more public housing units. The impermanence of the boundaries, however, leave Town Councils relatively little time to implement big or substantial changes. Residents and residents' committees might also find it more difficult to develop "town" identity and the sense of ownership of their estates which can contribute towards increasing participation in the management of these estates.

Conclusion

Only the future will show whether Town Councils can provide for public housing estates a new management approach different from the standardisation and uniformity that were characteristic when the HDB was managing the public housing estates. The physical changes that Town Councils are going to introduce are likely to be selective and complementary to the substantive character of public housing estates that has already been put in place by the HDB.

Nevertheless and in spite of the resource constraints, Town Councils are institutions that have been put in place to enable the residents to exercise their choices and preferences. Although they have to bear the costs of these choices and preferences, Town Councils are still intended to localise some of the decision-making on the management of public housing estates. In this sense, they have introduced a new and different phase to the public housing programme in Singapore.

The workings of Town Councils will also depend on the interest that residents show in self-determination as well as the participation in the running of their estates. So far, the interest seen among the residents other than those nominated to the councils and their various committees and sub-committees can best be described as lukewarm (Ooi 1990). Apart from the reasons already discussed above, this relative lack of interest in participating in the running of their estates can be partly explained by the limited role of Town Councils, that is, the narrowness of the range of tasks and powers that Town Councils have been given. In terms of its tasks and powers, Town Councils appear to be less like local authorities and more like the management committees of private condominium properties which have been formed to take care of their day-to-day operations.

Decision-making on public housing policies has been highly centralised for the last two decades or so. Repetitiveness, as observed earlier in the discussion, has been the backdrop to HDB's policies for the public housing estates with the selective introduction of variations, "for good effect", not because the residents wanted or needed them. Centralisation brought efficiency and effectiveness in the delivery of housing and its support services. The costs involved a high degree of uniformity not only in the appearance of estates but also their management regulations. There were few mechanisms which existed for the residents to exercise choice and options.

Town Councils have been introduced to allow for residents' participation in the day-to-day running of their estates. A wider political agenda was to promote greater voter accountability as well as residents' responsibility for meeting the costs of the upkeep for their estates. The process of public participation in decision-making on policies for public housing estates, as has been underscored, requires wider interest in such participation than what exists at present. In order to promote wider participation, Town Councils might have to consider modifying the existing framework of incorporating the residents' views and "organise" more participation. They have to accept what one councillor had expressed to be the inconvenience of having to consult in an organised and institutionalised manner with the residents, delay in arriving at a decision and implementation of the decision and ultimately, more work for Town Councils.

REFERENCES

Field, B. (1987) "Public Housing in Singapore", *Land Use Policy*, April, pp. 147–156.

Liu, T.K. (1975) "Designed for Better Living Conditions", in S.H.K. Yeh (ed.) *Public Housing in Singapore*, Singapore University Press for Housing and Development Board, (Singapore), pp. 117–184.

Liu, T.K. (1985) "Overview", in A.K. Wong and S.H.K. Yeh (eds.) *Housing a Nation — 25 Years of Public Housing in Singapore*, Maruzen Asia for Housing Development Board, (Singapore), pp. 1–29.

Master Plan: Reports of Study Groups and Working Parties (1955) Government Printing Office, (Singapore).

Ministry of National Development (1988) *Town Councils — Participating in Progress* (Singapore).

Ooi, G.L. (1990) "Town Councils in Singapore — Self-determination for Public Housing Estates", Occasional Paper No. 4, Times Academic Press for Institute of Policy Studies, (Singapore).

Ooi, G.L. (1991) "Public Housing Estates in Singapore — A New Spatial Order", *Solidarity*, Nos. 131–132, July–December, pp. 123–134.

Ooi, G.L., S. Siddique and Soh, K.C. (1994) *Management of Ethnic Relations in Public Housing Estates*, Times Academic Press for Institute of Policy Studies, (Singapore).

Phang, S.Y. (1992) *Housing Markets and Urban Transportation* — Economic Theory, Econometrics and Policy Analysis for Singapore (Singapore: McGraw-Hill).

Pugh, C. (1989) "The Political Economy of Public Housing", in K.S. Sandhu and P. Wheatley (eds.) *Management of Success — The Moulding of Modern Singapore* (Singapore: Institute of Southeast Asian Studies), pp. 833–859.

Tan, T.K.J., Loh, C.T., Tan, S.A., Lau, W.C. and K. Kwok (1985) "Physical Planning and Design", in A.K. Wong and S.H.K. Yeh (eds.) *Housing a Nation — 25 Years of Public Housing in Singapore,* Maruzen Asia for Housing Development Board, (Singapore), pp. 56–112.

4

The Built Environment and Transport

Phang Sock Yong

Relationship between the Built Environment and Transport

The built environment and transport are intimately linked. In the long run, changes in transport costs as well as transport policies and systems have implications for patterns of residential and employment locations. On the other hand, changes in land-use patterns alter the density as well as the spread of employment and residential location, thus exerting an enormous influence on the demand for transport. This interaction between land-use and transport are on-going and it is not always possible to distinguish between cause and effect.

Urban economics, which focuses on the spatial dimension of urban growth, has tended to treat transport as the exogenous variable. Generally, post-war changes in the built environment such as the "flattening" of urban employment and residential density gradients over time, have been due to, amongst other factors, the reduction in transport costs arising from the increasingly widespread use of the truck and car as well as the construction of highways. Differences in urban densities in the United States, Canada and Western Europe have been due to differences in the overall population densities of these countries as well as the large variations in the rates of taxation of car use and ownership (Pucher 1988, pp. 296–299).

A major mass transit system will have an immediate effect on transport mode choice and trip-making behaviour. In the longer term, sites near stations will become more desirable as places to live in and work at and this will result in significant changes to the urban form. Thus, transport policies affecting costs and accessibility have significant effects on the built environment, especially in the context of rapidly growing cities with significant new construction activities each year.

In contrast to urban economics, transport economics has tended to adopt the *short*-run assumption that the causal link is from land use to transport. Land use is generally assumed to be pre-determined in the modelling of transport demand and in the choice of the most efficient transport systems to meet such demand. For example, post-war low density and sprawling suburban developments in the United States has made it inefficient for most cities to be served with mass transit systems. Most West European countries and Canada have, however, either explicitly prohibited sprawling suburban development or strongly encourage high-density clustered development (Pucher 1988, p. 397); urbanised areas in these countries have thus remained compact enough to be served by mass transit systems.

Similarly, government policies on zoning, density, infrastructure development as well as urban renewal and housing subsidies would have an impact on land use and hence, transport demand, feasible transport systems and mode choices. In reality, therefore, either approach by itself is not entirely satisfactory. What is required, especially in the context of land use and transport planning in the dynamic and rapidly growing cities of Asia, is a recognition of the complex interaction between the built environment and transport. This chapter analyses the interactions between land use and transport policies in Singapore which have resulted in continuous modifications to the built environment.

Some Land Use and Transport Statistics

In 1990, Singapore had a population of 3.02 million people and a land area of 641.4 square kilometres. This makes it one of the most densely populated cities in the world with an average density of 5 000 persons per square kilometre. As a consequence of limited land resources, the Singapore government has adopted a highly interventionist approach with regard to both land use and transport.

Singapore has enjoyed an average annual growth rate of 6.5 per cent in its GNP per capita between 1965 and 1990. Although it has one of the highest per capita incomes in Asia (US$11 160 in 1990), only 31 per cent of households were car-owning households in 1990. In contrast to cities in the USA, Western Europe and Canada where between 3 and 26 per cent of all trips in urban areas are made by public transport,[1] in Singapore, public transport accounts for as high as 56 per cent of worktrips (see Table 4.1).

Table 4.1 Mode of Transport to Work, 1980 and 1990

Transport Mode	1980 (%)	1990 (%)
Car only	13.6	16.7
Bus Only	56.0	46.1
MRT/MRT & Bus	—	9.9
Others	13.0	15.4
Not Required	17.4	11.9

Note: 1990 data are based on 10% sample.

Source: Department of Statistics, Singapore (1991) *Census of Population 1990, Advanced Data Release*, p. 25.

The efficiency of the transport system in meeting the demands for the movement of both goods and people is an important determinant of Singapore's competitiveness as well as the quality of life. Rising incomes and urban development of Singapore during the past two decades have, therefore, been accompanied by massive public transit infrastructure investments as well as a host of policies affecting vehicle ownership and transportation mode choice. Transport policy has been accorded much attention by the government during the past 20 years. Various innovative though tough policies such as the Area Licensing Scheme, taxes on car ownership and usage, a Motor Vehicle Quota Scheme, a Weekend Car Scheme as well as large infrastructure investments have resulted in relatively congestion-free roads in Singapore.

Peak-hour traveling speed in the central business district averages 30 kilometre per hour which compares favourably with peak-hour speeds ranging from 10 to 12 kilometre per hour in New York, Manila, Lagos, Calcutta and London. However, the preoccupation of planners has not simply been on mobility *per se*. Instead, land-use planning to minimise the need to travel has been an integral part of Singapore's long term use-transport strategy. The ratio of investment of Gross Domestic Product on transport averaged 19, 27 and 36 per cent in the 1960s, 1970s and 1980s respectively. During the past three decades, the proportion of investment in construction and works as a proportion of total investment averaged 54, 38 and 57 per cent respectively[2] — making Singapore one of the most newly built cities in the world. This rapid rate of construction and urban development has been accompanied by significant changes in the land-use pattern.

Since the 1960s, the dominant force behind construction activity in Singapore has been the public rather than the private sector. Through the Land Acquisition Act which was enacted in 1965, the Singapore government has become the largest landowner in Singapore. Government agencies and statutory boards such as the Urban Redevelopment Authority, Jurong Town Corporation and the Port of Singapore Authority control large tracts of commercial and industrial land and determine to a large extent, the location of employment. Public housing, which housed eight per cent of the population in 1960 and 87 per cent of the population in 1990, is built by the Housing and Development Board which effectively determines the location of the majority of residential units. In the following section, the activities of these institutions which determine to a large extent the pattern of demand for transportation will be briefly described.

The 1960s: The Genesis of New Institutions

During the 1960s, the Singapore government was understandably more concerned with the pressing problems of housing provision and employment creation. The majority of the population resided in congested and dilapidated housing in and close to the central city area. The unemployment rate in

1960 was about 13.5 per cent, with most jobs confined to the core of the city. Under these circumstances, the government accorded top priority to the task of employment creation and public housing construction. Public transportation took a "back seat", so to speak, and remained in the hands of the British-owned Singapore Traction Company Limited and the private sector, specifically 11 Chinese bus companies.

At that time, the general standard of bus services was poor; speeds were low, schedules unreliable, and convenience and comfort minimal. However, per capita incomes were very low as well.[3]

The Housing and Development Board (HDB) began operations in 1960, replacing the Singapore Improvement Trust which had been set up by the colonial government in 1927. During the 1960s, the HDB built more than 100 000 units of public housing in new towns which were located primarily on the fringe of the central urban area. These new towns include Kallang, Queenstown and Toa Payoh. The Economic Development Board (EDB), set up in 1961, was responsible for the development of industrial estates of which the Jurong industrial estate (a sprawling site on the south-west) was the largest. Smaller industrial parks were also established in and close to public housing estates and new towns. In 1968, the task of developing and managing industrial estates was transferred from the EDB to the Jurong Town Corporation (JTC).

Public housing in new towns, together with service sector and traditional employment which remained centralised, and the location of a major industrial estate in an outlying and sparsely populated area resulted in an increased need to commute longer distances. Increased commuting, rising incomes as well as the inadequacy of public transportation caused private automobile and motorcycle ownership rates to increase rapidly during the 1960s. There were 70 000 private motor cars in 1961 and by 1970, the number of motor cars more than doubled to 143 000.[4] Over the same period, the number of public motor cars (mainly taxis) and buses increased by 64 per cent. Assuming that the degree of utilisation of public motor cars and buses is approximately six times that of private motor cars, the increase in demand for road space by these three categories of vehicles alone was about **114 per cent** for the decade.

However, the total route kilometres of public roads increased by only 35 per cent over the same period. Moreover, local and collector roads have much smaller capacities and are, for the most part, rarely taxed even during peak demand periods. The expansion of effective highway capacity is dependent on the provision of major arterial roads and expressways which are geared to meet flow requirements. The construction of major arterial roads during the 1960s, however, lagged far behind the increase in motor vehicles. There existed 214 kilometres of major arterial roads in 1961 and by 1970, the figure was 240 kilometres, a meagre increase of only 26 kilometres or **12 per cent**. The number of private cars per kilometre of arterial roads increased from 328 in 1961 to 594 in 1970.

The result was serious traffic congestion, especially in the central area. Not surprisingly, the need for proper land use and public transportation planning grew. In 1967, the government initiated the preparation of a Concept Plan for purposes of long-range land-use planning. In 1968, the Ministry of Communications was formed and a transport advisory committee appointed to review the public transport system. The Ring Concept Plan was selected in 1971; the plan included provisions for the intensification and expansion of the central area improvements in transport via a network of expressways and Mass Rapid Transit System (MRT) as well as industrial centres and other workplaces located close to, or within easy reach of, public transport and residential areas.

The 1970s: Area Licensing Scheme and Measures to Restrain Level of Car Ownership and Use

The necessity for a large investment in an MRT system, as provided for by the Concept Plan adopted in 1971, became a subject of much debate throughout the 1970s and early 1980s. Even as the subject was being debated, the 1970s witnessed further decentralisation of the population as three major new towns (Clementi, Bedok and Ang Mo Kio), housing a total of more than half a million people, were developed in the outlying areas.

At the same time, rent de-control in the Golden Shoe area of the Central Business District (CBD) as well as the Urban Redevelopment Authority (URA) Sale of Sites programme to facilitate urban renewal[5] resulted in the rapid growth of central area employment density. While employment remained fairly centralised, the massive public housing programme resulted in rapid population decentralisation. This apparent locational mismatch of employment and housing further increased the demand for transport. On the other hand, constraints on the supply side of transport limited the density of central area employment. A policy to encourage the use of public transport and to discourage both car ownership and usage was adopted.

The government's first attempts to improve public transportation services included efforts to improve the efficiency of the private bus companies. The bus companies were merged into four regional companies in 1971, and in 1973, the surviving three companies were merged to create the Singapore Bus Service (SBS) Ltd.

Incorporated in the strategy to alleviate central area congestion were mechanisms such as road taxes and registration fees for private cars which were progressively increased during the 1970s in order to discourage car ownership. The additional registration fee as a percentage of the market value of the car had been increased from 15 per cent in the early 1970s to 175 per cent by 1983.

In 1975, Singapore initiated a road pricing scheme — the Area Licensing Scheme (ALS) — to reduce congestion during the peak hours between 7.30 am and 10.15 am.[6] Private cars (which include taxis) carrying fewer than four persons entering the restricted central area were required to pay a fee. Buses, motorcycles and goods vehicles were exempted from paying any fees. The daily ALS fee for private cars was $3 from June to December 1975. It was raised to $4 in January 1976 and to $5 in March 1980. The fee was reduced to $3 in 1989 when the scheme was extended to the evening peak (4.30 pm to 6.30 pm), however, exemptions were removed for car pools, motorcycles and goods vehicles.[7] While increasing the monetary costs of travel for motorists to the CBD, the ALS reduces the time costs of travel for both public transport users as well as motorist who are willing to pay the entry fee.

The longer-term effect of the ALS on land use is to encourage firms and other urban activities, for which a central location is not imperative, to locate outside the zone. Studies conducted shortly after the implementation of the scheme have indicated that the ALS did not have any significant effect on retail sales, labour availability and commercial space costs and availability. The ALS has also had limited effect on land-use patterns as most land sales within the restricted zone are controlled by the URA. The ALS, by reducing peak-hour congestion in the restricted zone, has actually increased the attractiveness of the CBD as a financial and service centre.

The above restrictive measures on car ownership and usage, together with higher energy prices during the 1970s slowed down the growth of private car ownership. The number of persons per private motor car remained constant (at about 16) from 1974 to 1980. This was so despite substantial increase in per capita incomes and the rapid lateral expansion of the urban area during the same period. The number of private motor cars grew only seven per cent between 1974 and 1980, while there was rapid growth in other categories of vehicles.

A major programme of public roads and expressway construction and improvement was begun in 1979. Major arterial roads and expressways increased from 308 kilometres in 1979 to 535 kilometres in 1986, amounting to a 76 per cent[8] increase in capacity within a seven-year period!

Despite the "car restraint" policies and road construction programme mentioned above, further increases in central area employment density and workforce, were again constrained by central area road capacity by the late 1970s. Further decentralisation of employment would have been necessary and inevitable if an MRT system had not been built. The opinion of some consultants engaged for the evaluation of the MRT project was that there existed much scope for reducing travel demands by improving population and employment balance. Amongst the suggestions made were:

a) the development of business — commercial centres and hotel complexes near the Changi airport and in the Jurong Town Centre;
b) increase efforts by the HDB and JTC to provide employment opportunities in the New Towns and other outlying areas;
c) the relaxation of HDB policies which unnecessarily inhibit the mobility of residents and
d) greater coordination amongst the government agencies responsible for Singapore's urban development to ensure a more balanced mix of employment and residences so as to reduce transport demand (MRT Review Team 1980, pp. 143–147).

The perception amongst some government planners, however, was that the continued growth of Singapore as an international financial centre and a service centre of the region might hinge on the presence of an MRT system. It was feared that these services and other activities, for which the agglomeration economies associated with a central area location were extremely important, might opt not just to locate away from the central area but from Singapore altogether (Mohring 1983).

The 1980s: The Mass Rapid Transit Era

In May 1982, after a decade of intensive debates, the decision was finally made to begin building a $5 billion MRT system and actual construction of the system commenced in 1984. The Mass Rapid Transit Corporation (MRTC) was established as a statutory board to undertake the construction of the system. In August 1987, the government created a private company, the Singapore Mass Rapid Transit (SMRT) Limited, which is owned by the MRTC. The MRTC owns the MRT system and leases the operating assets to the SMRT which has the responsibility of running the system. Part of the MRT system began operations in November 1987, and by July 1990, the whole system, comprising 41 stations with a route of 67 kilometres was completed.

A private company, Transit Link, was set up in November 1987 to oversee the integration of the MRT with existing bus services through the use of a common bus-rail ticket. The MRT has had a profound impact on the spatial structure of land prices and consequently, on the location of jobs and residences. The MRT makes the CBD and the Orchard area more attractive for firms as these areas become more accessible to employees as well as shoppers. During the past decade, the construction of the MRT was accompanied by a surge of high density development in these two areas. At the same time, the MRT links several HDB new towns to the CBD, thus reducing the time cost of travelling. With this improvement, there is less aversion to residing in outlying new towns which are linked by the MRT to the CBD such as Tampines, Pasir Ris, Yishun and Jurong East.

Since the commencement of operation of the MRT system, several additional measures to discourage car usage as well as car ownership have been implemented. These were considered to be necessary as the number of persons per private motor car had decreased from 15.8 in 1980 to 10.0 in 1989. Measures include increases in road and petrol taxes, the extension of the ALS to the evening-peak (4.30 pm to 6.30 pm) on weekdays in 1989 and the whole day (7.30 am to 6.30 pm) from January 1994, as well as the implementation of a motor vehicle quota scheme in 1990.[9]

The motor vehicle quota scheme requires the buyer of a new vehicle to possess a certificate of entitlement (COE) which may be obtained in a sealed bid public tender held monthly by the Registry of Vehicles. The COE is valid for a period of ten years and owners of cars older than ten years must pay the prevailing quota premium (which is the average of quota premiums over the past year). This scheme has allowed the government to effectively control the number of motor vehicles in Singapore.

In 1991, a Weekend Car Scheme[10] which restricts car usage for weekend cars was implemented. In return, owners enjoy tax rebates on the registration fee, import duty and road taxes. Weekend cars may only be used on Sundays and public holidays and during off-peak hours (between 7 pm and 7 am on weekdays, and after 3pm on Saturdays). A weekend car is identified by a red number plate and a day license is required to be displayed if the car is used outside the permitted hours. The Weekend Car Scheme as well as the Area Licensing Scheme, both of which represent crude road pricing schemes, are expected to be superseded by an Electronic Road Pricing Scheme in 1996.

The Future: The New Concept Plan

More significantly, the MRT's profound impact on the urban structure has been the impetus behind the review of the Concept Plan which began in 1987 and which was completed in 1991 (URA 1991). The new Concept Plan, building on the previous plan, focuses on both the development of Singapore as an international hub city as well as improving the quality of life of Singaporeans. The Plan addresses land-use planning in three stages: up to year 2000, year 2010 and year X when Singapore's population is expected to stabilise at four million.

Plans for extensive decentralisation of commercial and retail activities to four self-contained regional centres (see Chapter 1) constitute the cornerstone of the Concept Plan. This aspect of the plan has, however, been the subject of much public debate. It could be said that the controversy dates back to the public transport options debate on the construction of the MRT system in the early 1980s. Proponents of the MRT system had then argued that decentralisation of employment, as an alternative option, would not constitute a viable solution to the central area traffic congestion problem since Asian cities, unlike Western ones, were still experiencing strong centralisation trends. The MRT system was eventually considered by decision-makers to be a necessity in order, partly, for the central area to accommodate a larger workforce.

However, extensive decentralisation of employment in the future need not necessarily be inconsistent with an excellent transport system. Complementarity is the keynote here. Specific industries such as finance, insurance, commodity trading and international hubbing have strong agglomeration economies and as such, have traditionally been found in the downtown areas. By year X, commercial floor space in the downtown area (which will comprise the present CBD and the area around Marina Bay) is expected to be in the region of 14 million square metres — double the present downtown commercial space of 7 million square metres in 1991. The projected availability of space and ease of commuting to the central area via a vastly improved transport system will further reinforce its attraction for firms in the clusters mentioned above.

The four regional centres in Jurong East, Woodlands, Tampines and Seletar are projected to ultimately have 1.5 million square metres of commercial floor space each. Six sub-regional centres will each supply 0.5 million square metres of commercial floor space. The regional centres are, therefore, expected to be much smaller than the downtown area. This projected decentralisation trend for commercial space is, by no means, excessive when compared to Western cities, and especially when one considers the dispersed spatial structure of the population as a result of the location of the HDB new towns.

The New Concept Plan also provides for improvements to the transportation network via the construction of new expressways, new MRT links, a light rail system as well as an improved ferry system. Construction work on the Woodlands extension is in progress and is expected to be completed by 1996; by year 2010, the north-east line serving areas like Serangoon, Hougang and Ponggol is expected to be in operation. Careful land-use planning and the de-regulation of land and housing markets, in order that workers may reside near their workplace and thus, reduce the need to commute long distances, may yet be the most efficient as well as equitable land use-transport policy in the long run.

ENDNOTES

1 See Pucher (1988), pp. 377–402. As reported by Pucher (based on 1978 to 1981 data), transit's share of urban modal split was three per cent in the United States, 11 per cent in West Germany, 15 per cent in Canada, 19 per cent in Great Britain, 20 per cent in Switzerland and 26 per cent in Italy.
2 Investment refers to Gross Fixed Capital Formation. Averages are calculated from the Deparment of Statistics, Singapore (various years) *Yearbook of Statistics, Singapore.*
3 The per capita indigenous GNP was S$2 862 in 1966.
4 Figures are derived from transport data in the *Yearbook of Statistics, supra,* note 2.
5 Urban Redevelopment Authority (1989).
6 See Holland and Watson (1978) and Wilson (1988) for evaluations of the scheme.
7 Monthly fees are 20 times the daily fees and company cars pay double the rates of private cars. Motorcycles pay $1 or $20 monthly.

8 This figure was derived utilising figures for arterial roads and expressways derived from the *Yearbook of Statistics*, *supra*, note 2 and Planning Department, Singapore (1983) *Land and Building Use, Report of Survey*.
9 See Phang (1992), Chapter 8, for details on and analysis of the motor vehicle system as well as the scrap car system.
10 See Phang (1993) for details.

REFERENCES

Department of Statistics, Singapore (various years) *Yearbook of Statistics, Singapore.*

Department of Statistics, Singapore (1991) *Census of Population 1990, Advanced Data Release*, p. 25.

Holland, E.P. and P.L. Watson (1978) *Relieving Traffic Congestion — The Singapore Area License Scheme*, World Bank Staff Working Paper No. 281.

Mohring, H. (1983) "The Singapore MRT: What Price a Large Central Area Workforce", *Suara Ekonomi*, Vol. 20, pp. 9–15.

MRT Review Team (1980) *Singapore's Transport and Urban Development Options — Final Report of the MRT Review Team* (Singapore).

Phang, S.Y. (1992) *Housing Markets and Urban Transportation — Economic Theory, Econometrics and Policy Analysis for Singapore* (Singapore: McGraw-Hill).

Phang, S.Y. (1993) "Singapore's Motor Vehicle Policy — Review of Recent Changes and a Suggested Alternative", *Transportation Research A*, Vol. 27A, No. 4, July, pp. 329–336.

Planning Department, Singapore (1983) *Land and Building Use, Report of Survey.*

Pucher, J. (1988) "Urban Public Transport Subsidies in Western Europe and North America", *Transportation Quarterly*, Vol. 42, No. 3.

Urban Redevelopment Authority (1989) *The Golden Shoe*, (Singapore).

Urban Redevelopment Authority (1991) *Living the Next Lap — Towards a Tropical City of Excellence* (Singapore).

Wilson, P.W. (1988) "Welfare Effects of Congestion Pricing in Singapore", *Transportation* , Vol. 15, No. 3.

5

The Political Economy of the Built Environment Revisited

Linda Low

Introduction

The Singapore economy generally operates on free market principles but a very significant resource which has no real market is that for land. Although there are private residential and non-residential properties which are transacted in the open market, this private real estate market is very small compared to the amount of land and properties owned by the state. In 1949, state land accounted for 31.0 per cent of the main island, increasing to 44.0 per cent in 1960, 49.2 per cent in 1965, 65.0 per cent in 1975, 70.0 per cent in 1980 and 76.0 per cent in 1985.[1] Though the percentage of the population housed in owner-occupied Housing and Development Board (HDB) flats has increased from 37.0 per cent in 1970 to 42.0 per cent in 1980 and 87.0 per cent in 1993, these are not freehold properties but leased out for 99 years only.[2]

While provisions in the State Lands Act and related legislation have contributed to this growth, the government's powers of acquisition have a larger impact on the rise in state landholding.[3] The Land Acquisition Act under which land may be acquired for a certain public purpose, further confers powers to 11 other statutes, including those of the HDB, Port of Singapore Authority (PSA), Jurong Town Corporation (JTC), Urban Redevelopment Authority (URA), Economic Development Board (EDB), Public Utilities Board (PUB) and Telecommunication Authority of Singapore, Mass Rapid Transit Corporation (MRTC) and National

University of Singapore (NUS) for their infrastructure and estate development.[4] Between 1949 and 1984, the agencies acquired land including the HDB, JTC, URA, PSA, and Land Office (*Straits Times*, 18 October 1986). These acquisitions give a total of 81 243 hectares which make up about 30.2 per cent of the total land area of Singapore.

The infrastructural investment of these statutory boards, various laws and government policies affecting land allocation and utilisation, all have a tremendous impact on employment, housing location and commuter decisions in the context of a city-state. Land use in an urban setting as defined under a Master Plan is controversial (Gamer 1972 and Leung 1987). The 1955 Master Plan, adopted in 1958, is reviewed every five years from 1965 onwards. This plan delineates land use by "zone" designation, density ratio and plot ratio prescriptions. It regulates land values and levies a development charge. Physical development projects of the government and statutory boards are exempted from the Master Plan.

The built environment in Singapore has tremendous economic as well as socio-political impact. This warrants a political economy approach to study the impact and implications of such a land utilisation and ownership pattern. The main objective of this paper is to analyse issues arising from such a pattern of ownership and control of the built environment, and the prospects of such policies. First, a discussion of the concept of political economy as is understood in this paper for Singapore, is germane.[5] This is followed by a brief overview of the pattern of the built environment to tease out the issues to be discussed before concluding with some prospects and policy suggestions.

Political Economy of Development in Singapore

The unique economic philosophy of economic planning and macro-economic management in Singapore has been discussed elsewhere (Low et al. 1993). In the 1960s, when the People's Action Party (PAP) formed the first self-independent government, it had a tremendous uphill task faced with poor economic prospects. Singapore's mainstay entrepot trade had stagnated and there was socio-political unrest due to Communist-inspired conflicts in industrial relations. The political economy approach of the PAP government was thus grounded in the premise that, to restructure the economy and turn Singapore around into a viable nation especially after its exit from Malaysia in 1965, a holistic and determined effort required a strong interventionist government. That the PAP espoused socialistic principles in its early days with a strong sense of paternalism further defined its authoritarian style.

Being in continuous power since 1959, the PAP and Singapore government has become as synonymous as two sides of the same coin (Lim et al. 1986). While the PAP has lost some political ground in recent general elections as the political process filters back the dissent and unpopularity of some policies, the PAP leaders profess little compromise in making the necessary and core choices for Singapore even when they prove unpopular.[6] However, by the late 1980s and early 1990s, the PAP government has begun a process of rolling back the state in terms of scaling down the public sector, especially in terms of privatising its many government-linked companies (GLCs) and statutory boards. Policy reforms, tantamounting to a more accommodative "back to market" philosophy became evident as the private sector matured and was able to assume its role as the engine of growth in the economy (Low 1994).

Given the standard theoretical expositions on political economy and the peculiar features of the Singapore economy, the main thrust of the political economy approach is deemed as proposing policy changes in the form of hypotheses, usually of the leaders, to be subjected to the empirical test of political approval. A second trait of the political economy approach in Singapore is that, economic efficiency in resource allocation must be upheld. It must, however, be noted that there is a distinction between efficiency in the processes through which transactions, ideally under voluntary contracts, are carried out and efficiency in the results. This dichotomy may appear tautological. Nonetheless, reforms designed to enhance efficiency should apply to the institutional structure.

In the bid to attain efficiency, the line dividing individual choices and social outcomes may be blurred. Rather than try to find optimal solutions to economic problems, policy-makers should concentrate on finding good decision-rules which all individuals and interest groups would find it in their interest in the long run to adopt. Examples of such decision-rules are manifest in the land acquisition, public housing, population and land transportation policies. The PAP leitmotif is that the survival of the society, as a whole, precedes the rights of individuals who make up the whole. Paternalism enters the decision-making process.

With the above perception of the political economy in Singapore, three problems can be perceived to plague the traditional economic approach to politics, namely, the utilitarian calculus, the social engineering urge and the elitist mentality. The first difficulty is the recognition that a social welfare function cannot, in general, be derived from individual preferences by majority vote or any aggregation process in such a way as to represent collective preferences. This does not, however, mean that such a function is not a useful tool for analysing the fundamental trade-off between efficiency and social justice in the analysis of economic policy.

The problems of social engineering and elitist mentality are urged on by economists who believe, they know something about optimal social economics and social order, that the participants themselves do not. An elitist mentality is, perhaps, the most complex of the three, with the dangers of intellectual and academic arrogance. The latest PAP slogan: "more good years ahead" – to show the way towards the good society and be a developed industrial country by year 2000, demands a compromise between some elitism in policy-making and implementation and a more participatory and consultative style of government.

Scope and Pattern of the Built Environment

The pattern of land use in Singapore since 1950 has been summed up and shown in Table 5.1. Over the period 1950 to 1993, share of built area has grown from 18.5 per cent to 48.6 per cent, a more than two-fold increase in absolute magnitude. Intensive utilisation of scarce land and high opportunity costs have reduced the share of land occupied by farms, forests and marsh and tidal wastes. The amount of land devoted to other uses has grown.

Table 5.1 Land Use, 1950–1993

Year	Total	Built-up Areas*	Farm holding Areas†	Forest	Marsh & tidal waste	Others#
km²						
1950	755.2	139.4	181.4	197.0	47.4	190.0
1970	586.4	189.9	134:0	32.4	32.4	197.7
1993	641.4	316.5	6.9	28.6	15.7	273.8
% distribution						
1950	100.0	18.5	24.0	26.1	6.4	25.1
1970	100.0	32.4	22.9	5.5	5.5	33.7
1993	100.0	48.6	3.3	4.5	2.4	42.0

* includes new industrial sites.
† refers to licensed farms exclude land occupied by pure rubber & coconut plantations.
\# includes inland water, open spaces, public gardens, non-built up areas in military establishments, quarries, rubber & coconut plantations.

Source: Department of Statistics, Singapore (various years) *Yearbook of Statistics*, Singapore.

There is no further published information on the allocation of land within the built-up categories between the private and public sectors by use. A rough estimate of the state owning 80 per cent of total land by 1990, may not be too far wrong even if six to seven per cent of this comprise forest and marsh and tidal waste. On the other hand, the government has the resources and power to reclaim more land from the sea. Reclamation has added about ten per cent to the land area since 1967 though five per cent has been allocated for reservoirs. But reservoirs are crucial to Singapore which depends on Johor in Malaysia for some 50 per cent of its water needs and is trying to diversify to Indonesian sources in Bintan. Under the current plans which involve more sophisticated technology to go deeper into the seabed, Singapore expects to reclaim another 26 square kilometres or 4.1 per cent of total land area.

Overall land use proposed in year X when the population reaches four million compared with the pattern in 1991 is shown in Table 5.2. Land for institutions is meant, inclusively, for schools, hospitals and other institutions. Recreation includes all open space while infrastructure is inclusive of utilities and telecommunication. Transportation includes roads, ports, airport and the MRT. "Others" as a land-use category is inclusive of cemetry, agriculture, catchment, reservoirs and others which will drop drastically from 47.0 per cent of land use in 1991 to 22.0 per cent in year X. All categories of land use increase in year X, especially the doubling in shares for residential use.

Table 5.2 Broad Land Use Proposed for Year X (population 4 million)

Land Use	1991 (pop 3 m)	Year X (pop 4 m)
Residential	11.5	20.5
Commercial	1.5	2.0
Industry	10.5	18.5
Institution	4.5	7.5
Recreation	10.0	11.0
Infrastructure	2.0	3.5
Transportation	13.0	15.5
Others	47.0	22.0

Source: Urban Redevelopment Authority (1993/94) *Annual Report*.

To visualise the impact of state land used for housing, Table 5.3 on houses and households by type of housing in the census years of 1970, 1980 and 1990, shows a distinct rise of public flats relative to private housing. The data for 1970 and 1980 are for a number of houses which, unfortunately, cannot be easily updated in 1990. Instead, we report the number of households living in various types of houses on the presumption that these figures would be a close approximate of such types of housing as well. Table 5.4 shows the total private households and resident population by type of dwelling in 1990.

What is most significant in Tables 5.3 and 5.4 is that public housing has overtaken private housing and that more land has been allocated to the public sector for public housing. This has to be so if the government wants to promote a 100 per cent home ownership policy and ensure a home for every Singaporean, especially the low-income households (*Straits Times*, 26 May 1994).

Table 5.3 No. of Houses by Type of House, 1970 and 1980 and Households by Type of Houses, 1990

Type of House	1970 No.	1970 %	1980 No.	1980 %	1990* No.	1990* %
Total	467142	100.0	305833	100.0	744203	100.0
Bungalows	9962	2.1	10711	3.5	11005	1.5
Semi-detached	12114	2.6	7465	2.5	16866	2.3
Terrace houses	21462	4.6	21500	7.0	26976	3.6
Private flats	16321	3.5	10008	3.3	44064	5.9
Shophouses	13453	2.9	19806	6.5	12985	1.7
Attap/zinc-roofed	50313	10.8	104188	34.0	6854	0.9
Public flats	337198	72.2	120138	39.3	612620	82.3
Others	6319	1.3	12017	3.9	12833	1.7

* households living in such housing types

Source: Yeh (1984), Table 5.1, p. 66 and Department of Statistics, Singapore (1991) *Census of Population 1990, Households and Housing.*

Table 5.4 Total Private Households and Resident Population by Type of Dwelling, 1990

Type of Dwelling	Households No.	Households %	Resident Population No.	Resident Population %
Landed properties	54847	7.4	238227	7.9
Bungalows	11005	1.5	49376	1.6
Semi-detached	16866	2.3	75721	2.5
Terrace	26976	3.6	113130	3.8
HDB/JTC	599810	80.6	2432543	81.2
1-, 2-rm	60367	8.1	181157	6.0
3-rm	252879	34.0	1009238	33.7
4-rm	190036	25.5	836701	27.9
5-rm, exec	90689	12.2	386754	12.9
Others	5839	0.8	18693	0.6
Other pub flats	12810	1.7	46883	1.6
Conds & pte	44064	5.9	147212	4.9
Shophouses	12985	1.7	41294	1.4
Attap/zinc	6854	0.9	33329	1.1
Others	12833	1.7	57738	1.9
Total	744203	100.0	2997226	100.0

Source: Department of Statistics, Singapore (1991) *Census of Population 1990, Households and Housing.*

State control of land and hence, its price instead of a free land market policy in Singapore is, perhaps, no different from that in other capitalist countries like France, the United Kingdom or Switzerland.[7] Land has to be publicly owned or leased to the public for 99 years, equivalent to "*superficy right*" in Switzerland or "*emphyteotic lease*" in France and the United Kingdom. These European countries have similar laws to empower them to acquire land cheaply in development areas as well as enjoy loans at below market interest rate or provide fiscal incentives for private developers interested to build social housing.

As a matter of opinion, the discrepancy between a real estate market controlled by the public sector and a wholly private one could be corrected through a system of long term lease of the land to the private sector. The price will be different from location to location according to central or specialised locations. It could represent a source of greater income for state coffers. While the variety of prices (in comparison with a completely free market economy) would be greater, land is ultimately under state control.

Moreover, the successful public housing policy in Singapore realised through state agencies means that complex legislation is not necessary as all the "actors" are under the state. If the private sector were dominant, there may arise a situation of over-legislation to control and achieve the desired outcome. Such over-legislations may create a paralysis and conflict between actors in the public and private sectors as evidenced in many Western countries.

Since the HDB and the URA are the statutory boards which affect the built environment most, an analysis of their activities is useful. The mission statement of the HDB, as the national housing authority, is that it provides a high standard of affordable housing and supports the building of communities. The URA's is to plan and facilitate with vision, creativity, efficiency and highest quality service, the physical development of Singapore into a tropical city of excellence with a distinctive identity in its effort at meeting the aspirations of Singaporeans.

Table 5.5 shows the quantum of modern facilities generated by sales of land by the URA since 1989/90.

Table 5.5 Modern Facilities Generated by URA Land Sales

Land Use	1989/90	1990/91	1991/92	1992/93	1993/94
Residential (units)	—	—	508	767	847
Commercial (m^2)	131392	11760	16627	23392	37183
Hotel (rooms)	520	—	—	345	736
Shophouses (units)	7	14	29	45	101
Heavy veh park (lots)	—	—	321	545	927
Industrial (m^2)	—	—	—	324528	—

Source: Urban Redevelopment Authority (1993/94) *Annual Report.*

Under a strategic approach in planning Singapore's physical development, the URA uses blueprints including the Concept Plan, the Development Guide Plans and the Conservation Master Plan (see Chapter 2). Studies are regularly made to ensure that the Concept Plan keeps pace with changing needs in housing, leisure, transport, community facilities and businesses. Innovative concepts like cluster housing and strata bungalows which are landed properties with shared facilities were introduced in 1993 for instance (see Chapter 1).

The Development Guide Plans give greater planning details for designated areas or neighbourhoods. These plans have been farmed out to the private sector to harness ideas and involve its participation. As for conservation, the URA has become the executive arm of the Preservation of Monuments Board (PMB) in 1993.

It is irrefutable that the URA has played as significant a role in developing facilities for an urban industrial country, as the HDB has, in public housing. The potent powers of these and other statutory bodies together with land-use policies implemented have largely gone unquestioned. This has been due to the presumption that they were given the task of national development such as breaking the back of the housing problem, redeveloping the urban area and providing the infrastructure to enhance Singapore's comparative advantage as a modern industrial and business hub.

The activities of the HDB are shown in Tables 5.6 and 5.7. The slowdown in some years is more of a conscious policy in response to the glut in supply or to upgrading of facilities.[8] Hence, lower supply of industrial premises, one- and two-room flats and the shorter rental queue are witnessed.

Table 5.6 Principal Statistics of HDB

Percentage / Units	1987/88	1993/94
% of population in HDB flats	86	86
Waiting list to buy (excl resale flats)	64271	76759
Waiting list to rent	3272	7281
Flats sold under HOS & HUDC schemes	30929	20925
Flats completed	29840	22023
Resettlement cases cleared	9919	3002

Source: Housing Development Board (1993/94) *Annual Report.*

Table 5.7 Residential Units under HDB Management

Type	1984	1994
1-room	60646	27446
2-room	48562	36644
3-room	209714	244280
4-room	87619	217589
5-room	32728	92437
Executive	2333	36989
Multi-generation	0	367
HUDC	3214	5411
Others	61	53
Total	444877	661216

Source: Housing Development Board (1993/94) *Annual Report.*

Both the HDB and the URA, especially the latter, are involved in the resettlement of residents and businesses. These are instances where the powers of the state to maximise societal welfare become potent. Resettlement under the URA is also the result of the de-control of premises whose values have been fixed under the Rent Control Act. The role of the URA in de-controlling premises is to allow for greater economic development by enabling landlords to pay reasonable compensation to sitting tenants, leaving acquisition as a last resort.

Issues in Political Economy of Built Environment

Some general issues concerning the built environment will be analysed first before going into the micro aspects affecting choices in location of industrial sectors and consequently, housing, transport, employment and other activities. As noted earlier, Singapore appears to have no market for land because the government is a major landlord with an estimated 80 per cent of the total land available in 1990. While the URA has made land sales to the private sector by tender, a substantial amount of land allocated to public housing and other infrastructural development is not priced under a market system.

The current pattern of land utilisation constitutes a long-term decision leaving few inter-temporal choices for future generations. Redevelopment within the confines of the Master Plan which is revised every five years would be immutable. In other words, the economic and socio-political decisions made in this generation would set the pattern and pace for future development.

The pattern of land use reflects the economic and social paths deemed appropriate by some collective decision process. The current plan is to be a high-technology industrial economy by year 2000 with a standard and quality of life equivalent to Switzerland to go with it. Accordingly, land has to be allocated to satisfy economic growth as well as more aesthetic goals involving conservation, environmental protection and upgrading the quality of life. Plans for more land for nature conservation have been submitted to the government by the Malayan Nature Society in Singapore to preserve the habitat of wildlife (Briffet

1990). This could have recreational, educational, scientific and even economic benefits.

Land set aside for such social and recreational needs also meet those of the tourist industry. With land scarcity, imaginative use of land under related multi-purposes is necessary. Hence, inland water serves as reservoirs, public parks, sites for cultural and sports activities. Even the luxury of cemeteries has to be given up except for certain religions which require interment.

Public-Private Sector Mix

While it may be argued that such a dominance of state ownership is justifiable and it would take a long time to lower state-land holdings, there is merit in considering a more imaginative land use and ways to attract private investment in order to better utilise land in terms of optimality and efficiency. For instance, some land is being sold in the HDB estates to private developers to integrate public and private efforts more tangibly. This will diffuse the current physical segregation of the two housing markets.

The URA has allowed two new housing forms, cluster housing and strata bungalows which can be built immediately, giving Singaporeans a wider range of choice in housing types (*Straits Times*, 6 March 1993). This also gives architects more room to exercise their design creativity. But, at the same time, to safeguard the character of the estates, the URA has identified 40 800 private landed residential properties out of a landed housing stock of 62 000 which will be protected from non-conforming building types (*Straits Times*, 14 December 1994). In addition, another 2430 units in 39 areas have been protected under the Good Class Bungalow Areas scheme since 1980. The 40 800 units will increase to 53 300 units in the future. Landed properties constitute 7.5 per cent of the total housing stock which will increase to 10.8 per cent or 165 000 units when Singapore's population reaches four million under the URA's Concept Plan.

With freehold land being depleted, the government through the URA and the HDB will sell land for 99-year leasehold tenure under reserve sites (*Straits Times*, 16 June 1993). The mentality of Singaporeans towards 999-year leasehold or freehold land will have to change and the government will release more land for 99-year leasehold tenure private housing for an average of 1300 properties a year to ensure supply to stay ahead of demand. These reserve sites are, in addition, to the Ministry of National Development's (MND) land sale programme for the building of 3000 private houses a year (*Straits Times*, 5 June 1993).

Being a small economy lacking in resources with limits on land resources, the argument for state ownership is cogent. However, there are implications of not pricing land at market rates because it is not regularly transacted freely like other commodities under the usual demand and supply paradigm. Such implications must be noted and ameliorated. The issue of land cost being subsidised for public housing or further port development is a good case in point. These will be discussed later. Such a concern has arisen now, probably because the socialist model of siphoning land to the state for infrastructure and promoting home ownership has achieved its limits and a capitalist model is now deemed desirable to reflect costs and benefits more efficiently and realistically.

The government has become the major stakeholder in land because of its careful planning of development paths. Such development-oriented strategies were aided by both the independent city status and PAP hegemony. For instance, under the First Development Plan which spanned the period 1960 to 1965, massive land resources were siphoned to the industrialisation programme and Jurong was earmarked for the manufacturing sector (Toh and Low 1988). Similarly, to solve the acute housing problem and other attendant problems involving environmental health and squatter colonies, the HDB was given state land freely. When the need to create some light industries in public housing estates arose, other satellite industrial estates sprouted up in the right places. Such land allocation would clearly affect other choices of housing location and commuter patterns, among other things (Phang 1989; Tan and Phang 1991).

Much controversy has arisen over the Land Acquisition Act and compensation rates. This issue appears to have been addressed by the amended Act in 1988 which changed the pegged price of the property acquired to its market value on 1 January 1986 or on the date of notice of acquisition, whichever is lower. However, the date picked for the new pegged rates was still one of the low prices as the property market had scarcely recovered from the 1985 crash then. There appears to be still some leverage for the government to acquire land at below market rates based on the argument that it is for the public good and positive externalities.

The Control of Rent Act dating from 1947 had hampered both private sector efforts to revitalise the Central Business District (CBD) and distorted prices as rents were capped. The Controlled Premises (Special Provisions) Act of 1969 facilitated urban renewal by the private sector within a designated area. *De facto* de-control through land acquisition and other measures such as the conservation programme, have been used within the CBD. In 1980, vacant domestic premises which had not been occupied since 24 October 1980 were exempted from rent control, constituting vacancy or "creeping" de-control. On 6 October 1988, the government announced that rent control would be phased out in stages by further selective implementation of block control under the Controlled Premises (Special Provisions) Act. Whereas 46 per cent of all buildings were affected by rent control in 1960, the percentage fell to 38 per cent in 1969 (Motha 1981, p. 147). Such de-control is in line with the government's move towards privatisation, to involve greater private participation, especially in the conservation of historical sites for tourism and posterity.

Affordable Homes

The most controversial issue in public housing is that of land subsidies. Essentially, the government plays "Robin Hood", acquiring land at sub-market rates to supply low-cost public housing. This has helped to set a more equitable pattern of income distribution and generally, kept costs low since housing is a large and basic item of household expenditure. While such a policy was justifiable in the 1960s and 1970s, it has become less acceptable with growing income levels. Some HDB flat dwellers are, reportedly, spending large sums on renovation and furnishings; as well on cars which are quite disproportionate to their subsidised housing.

Three sources of housing subsidy can be identified (Phang 1989). One is through annual government grants to cover the deficits of the HDB. Moreover, the HDB is able to obtain loans from the government at below the market interest rate. Finally, the power to acquire land cheaply by the HDB is another subsidy. While subsidisation may imply that the HDB has made some contribution towards income distribution and equity, this seems more apparent than real. Even without any rigorous cost-benefit analysis or use of more sophisticated measures like the Lorenz curve or Gini ratio to compare the state of income distribution before and after the successful public housing programme, some *a priori* evidence may be noted (Cheah 1989).

One is the observation that very conspicuous forms of consumption in luxury cars, furnishings and renovation are enjoyed by the HDB flat dwellers. Another is the squeeze on the middle-income group who neither qualify for subsidised HDB flats nor can afford the prices of private residential properties, leading to the formation of the Housing and Urban Development Company (HUDC) to meet their needs. Besides the inequality of welfare and utility among households living in public flats, one must not forget the inequality between public flat dwellers and those unsubsidised in private housing.

Since the income criteria of HDB flat buyers are screened only at the time of purchase, unless they wish to upgrade themselves to private housing which corresponds to their rising income over time, they enjoy the low-cost flats even after they have become better off. Such income ceilings have the objective of increasing the consumption of housing of the lower-income groups. However, an unintended side effect may be the "under-consumption" of housing by the higher-income groups.

Perhaps, partly in recognition of these trends of under-consumption, the government has encouraged homeowners to upgrade to larger and better residences. This also liberalises further the use of CPF savings. In 1993, $3509.4 million or 32.1 per cent of total withdrawals from the CPF were for approved housing in both the public and private sectors, approved non-residential properties and home protection insurance schemes. This amount in 1993 is very much dwarfed by the $6005.9 million or 54.9 per cent of total CPF funds withdrawn under the category of "others" — a dominant portion due to the Singapore Telecom shares offer (Low 1994).

The CPF balances form a large part of personal and household wealth and the rate of interest earned is less than what can be gained from investing in real property and approved stocks and shares which are also a better hedge against inflation. The policy to encourage upgrading may, however, be an overkill as speculative demand has been active and over-consumption of housing could also lead to misallocation of resources (Lim et al. 1986). From the official viewpoint, a larger home-owning population is also perceived as contributing to social and political stability and cohesion. The argument is that, to maintain healthy property values, the people must work to ensure and maintain economic and political viability.

The speculative gains from re-selling public housing were larger for flats bought before 1977 when prices were greatly subsidised. Such flats sold in 1982 were found to have returns of some 95 to 172 per cent (Phang 1989). This shows that either demand has shifted in favour of larger units as older flats are sold by those wishing to upgrade, or that larger units are subsidised more, hence the preference for them. Since 1979, with the relaxation of rules governing resale (which coincided with substantial price increases for new flats making the public housing market more market-oriented) and the property boom, profit margins for public housing have fallen. Resale levies progressively graded with size and other policy changes determining the resale prices, levies, occupancy and debarment periods (Wong and Yeh 1985, pp. 235–236) are also intended to lessen the profiteering. There is also a conscious policy to control resales of HDB flats to avoid the formation of ethnic enclaves.

On speculative trends and asset inflation, the asset bubble in 1993 reflects a shift away from prudent investment into a frenzy of buying property, stocks and shares and cars as both the economy and Singapore dollar were riding high. Like Japan, Singaporeans were under pressure to do something with their growing savings since interest rates remained low and in fact, turned negative in 1993.

The government's relaxation of CPF rules and the flotation of Singapore Telecom in 1993 further signalled the government's intent under its Asset Enhancement Programme. That the government has taken steps to cool the overheated property market by releasing more land, persuading banks to be more cautious in property and personal lending and generally, curbing credit card applications usage means that it is aiming for a soft landing to protect property owners.

This measure of preventing overheating, however, is not the central issue. More crucial is the government's promise to guarantee all Singaporeans, especially the low-income households, a home of their own (*Straits Times*, 26 May 1994). But somewhere along the many schemes and government intervention in the various components of the asset market is the phenomenon of asset inflation which is not a healthy trend. Between 1983 and 1993, private residential property values rose 82 per cent while car prices increased more than 200 per cent. From the trough which property values hit in 1986, private residential property values rose 156 per cent. While at the same time, monthly household incomes have increased, which means larger absolute loans and higher affordability, there is still the need for the government to intervene further from time to time to protect the genuine, especially first-time buyers. One may question whether the government has put too great an emphasis or set too accelerated a pace to its home ownership policy where a slower pace in tune with market forces might have been more desirable rather than an artificially created demand and speculation.

Even though the general aims of public housing upgrading or sale of shops and stalls are fundamentally correct, the impact on escalating costs have to be considered as the real estate market has been over-priced by induced demand and speculative activities. A more economic point is the amount of capital sunk into the residential market because of the government's housing policies. The amount of capital is obviously enormous and out of proportion to investment in machinery and equipment which is economically more productive (Ministry of Trade and Industry 1986, p. 6). With the attainment of dividends from the economics and politics of home ownership, it is time, perhaps, to play down such investment and land allocation to correct the balance.

Despite the market apprehensions discussed above, a very important feature in Singapore's land-use development must first be recognised — the financing aspect of the developments. Under the standard cost-benefit approach to undertake a public project, both costs and benefits span over generations, depending on the nature of the project. Some sharing of the financing burden would enter the picture, depending on whether the project is tax-financed or loan-financed, the latter whether domestic or external loans. Singapore is in a fortunate fiscal state of surplus where external loans are minimal and the domestic component is more from the Central Provident Fund's (CPF) obligatory holding of government securities. Even such public borrowing is not needed for financing development but ends up being invested abroad together with other public sector surpluses. The sale of shops, for instance, has increased HDB's surplus to $1.15 billion in the financial year 1993/94 compared to $288 million in the previous year (*Straits Times*, 26 January 1995). This came from selling 2048 shops compared to 851 shops a year ago.

In fact, the government's control of so much land and its ability to make revenue out of URA land sales gives it both the resources and ability to make good the promise of affordable homes to all Singaporeans. Four plans to help low-income HDB families to own HDB flats, benefitting about 4000 families were the result of brainstorming at the MND (*Straits Times*, 1 March 1995). Total government subsidies add up to $160 million a year with each family receiving about $35 000. On reflection that the HDB might have been conservative in expecting that families with monthly incomes not exceeding $1000 to own flats, it started to buy HDB flats in the open market and sold them at a discount to such families. It was pleasantly surprised that such families also have quite sizeable CPF savings. The HDB will, thus, analyse the financial backgrounds of applicants to find out if they can afford to pay more so that it could buy newer flats in better locations for resale under this scheme. So far, flats have been bought in the range of $80 000 to $110 000 and resold between $30 000 and $40 000. The selection rate for such flats, however, is 60 per cent compared to 90 per cent for new flats, that is, there is a smaller response rate to selection of the flats being resold. The HDB could look for flats costing up to $130 000. This scheme has attracted 2854 applicants by the end of 1994.

A second scheme involves the sale of rental flats to sitting tenants with a fourth batch comprising 235 upgraded two-room and three-room flats sold in 1995 at an average of $26 000 and $65 000 respectively. Tenants enjoy a three per cent discount for each year of tenancy up to a maximum of 30 per cent or $10 000, whichever is lower, and receive up to 100 per cent mortgage of the discounted price. The first two batches sold in 1994 comprised 511 three-room flats and 90 per cent were sold. The third batch of 317 two-room flats resulted in 72 per cent of the tenants buying. Among those who had decided against buying, many wanted to buy larger flats while elderly tenants decided to continue renting.

In 1993, budget four-room flats were offered for sale and up to January 1995, 2554 of such flats have been offered for sale. These flats are smaller than the standard four-room flats and have simpler fittings but cost 15 to 18 per cent less.

A fourth scheme which started in 1994 involve priority allocation to tenants in rental flats wanting to upgrade to three- or four-room flats. These applicants are given a six-month headstart over non-tenant applicants.

These schemes have been introduced, in addition to, another Small Families Improvement Scheme launched in 1993 which were designed to assist poor families that agree to keep their families small, that is, with two children, in order to upgrade. The government provides an annual education bursary from primary to pre-university or polytechnic education and an annual housing grant. The housing grant of $800 a year is paid into the wife's CPF account for 20 years or until she reaches 45, whichever is sooner. Up to $16 000 will be provided and the money can be used to buy a HDB flat. Both husband and wife must earn no more than $750 each month and have no "O" or "N" level passes, with at least one spouse a Singapore citizen or permanent resident.

Affordable homes which appreciate in value is part and parcel of the Asset Enhancement Programme for Singaporeans announced at the National Day Rally speech in 1993 by Prime Minister, Mr Goh Chok Tong. The Stall Ownership Scheme and encouragement to Singaporeans to own stocks and shares are other dimensions of asset enhancement.

In 1994, a number of schemes may be noted to fulfil economic, social and political goals. These schemes deserve mention because they are all, in one way or the other, tied up with the policy of public housing as an instrument to attain other objectives. Under the CPF Housing Grant Scheme announced in 1994, the government is using the housing policy to strengthen family ties by encouraging married children and their parents to live near each other. Under the scheme, a first-time applicant who buys a HDB resale flat in the same town or estate or within two kilometres of where his or her parents or children are staying and who satisfies all the eligibility conditions for a direct purchase flat will get a $30 000 grant credited to his or her CPF account. This can be used as a capital payment to reduce the principal sum of the mortgage loan. Applicants must be Singapore citizens or permanent residents who do not have a gross monthly income exceeding $8000 or own any private residential properties. A proper family nucleus is required as well as compliance with the approved neighbourhood and block ethnic limits.

Further, in 1994, a registration-for-flats system was launched to allocate flats in non-mature estates so as to achieve a fairer and more equitable balance between the needs of first-time applicants and upgraders. This scheme reflects the government's consciousness and sensitive monitoring of how the massive scale of its public housing policy may generate adverse reaction if not properly administered. Two separate queues exist for the groups identified above, with 60 per cent of available flat supply allocated to first-timers. This is an improvement over the 1989 booking system in which new flats were allocated each quarter by a computer balloting process.

In 1992, the HDB revised the allocation procedures for flats in non-mature estates to accord better chances to first-time applicants over upgraders who apply at the same time. This system worked well until the rapid increase in HDB resale prices and the very buoyant property market which led to a phenomenal increase in the number of flats applications. Upgraders accounted for the bulk of these applications and despite the priority given to first-timers, only a small number of them have an outside chance of getting a flat at their first try.

The HDB upgrading which programme which will span 20 years and cost the government between $10 and $15 billion, appears to be another scheme with obvious economic and political dividends. Older HDB estates will be brought up to the standard of newer towns and estates. This would ensure that residents in older estates will enjoy environments comparable to those in newer HDB and private housing estates without having to be uprooted. It will also ensure that they do not lose out in any asset and property appreciation. Since the government announced the plan in 1989, it has been cautious in beginning the pilot, demonstration and main phases as Singaporeans may find difficulty in visualising the programme.

Hence, the pilot phase which started in 1992 was followed by the demonstration phase in 1993 to allow residents to view models of upgraded precincts. Six precincts have been selected for the main phase in 1992 followed by another six in 1993. Thereafter, ten precincts will be listed for upgrading every year. Residents pay between 7 and 42 per cent of the upgrading costs, depending on the kind of flats and package opted for: basic, standard or standard-plus. The government's share of expenditure on upgrading works out to be in the range of $24 500 to $53 800, depending on the kind of flats and package at June 1992 costs. With upgrading, resale prices of upgraded three-room flats hit $160 000 in Kim Keat in the first month of 1995; twice the amount of the asking price in 1992 before residents voted for upgrading (*Straits Times*, 26 January 1995).

Supply of Land

To ensure that all Singaporeans can afford a home, the government has stepped up the rate of state land sales for landed property development. It used to release land for just 400 units of landed private properties in the early 1990s, increasing to 1100 in 1993 and 2300 in 1994 and 1995 as announced in 1993. By 1994, however, the number has increased to 3000 (*Straits Times*, 1 June 1993). The number of 3000 is about 50 per cent of the projected annual demand up to the end of the decade, that is, an annual demand of 6000 homes up to the year 2000. The sharp increase to 3000 reflects the government's concern with rising property prices. The URA's Private Property Price Index showed prices rose by 15.9 per cent in 1993 with the steepest increase for condominiums at 23 per cent.

However, while allowing individuals and small developers to bid for state land parcels, they will not be subsidised so that the "market" will be allowed to take its course (*Straits Times*, 18 September 1993). The government has also denied that a local property development magnate, Ng Teng Fong's organisation, Far East Organisation, is monopolising the private residential market as others such as the Hong Leong Group and its subsidiary City Developments Limited has 21.3 per cent of all private homes entering the market between 1993 and 2000 compared to Far East Organisation's 13.5 per cent (*Straits Times*, 18 September 1993). However, Far East Organisation has secured 39 per cent of state land parcels for development of private housing since the introduction of the land sales programme in 1989.

Land for Transport Policies

In the area of land transport which remains the most unsettled problem in Singapore, the issues of road congestion versus road expansion, rising affluence and modern commuting needs, remain (Low 1995). The 1970s had seen some success with fiscal policies — high road taxation and import duties — in curbing car ownership, coupled with high user charges such as parking fees and other motoring costs to discourage car usage. But with higher income levels from buoyant economic growth and generous government subsidisation in other social services like public housing, health and education, the income elasticity of demand for cars has been estimated to be 1.91 per cent over the period 1966 to 1987 (Chin and Koh 1989). This means that a one per cent rise in disposable income causes 1.91 per cent increase in car ownership.

In response, after the introduction of the certificates of entitlement (COEs) for the purchase of private vehicles and longer term measures like Electronic Road Pricing (ERP), a Land Transport Authority (LTA) was set up in 1995 to co-ordinate efforts to improve public transport system and manage traffic more effectively (*Business Times*, 15 March 1995). The government wanted to widen the choice of public transport services, raise service levels and allow more cars on the road without increasing congestion. The LTA will create a more comprehensive range of public and private transport modes using the Light Rail Transit (LTR), the Mass Rapid Transit (MRT), taxis and private cars. It will oversee these improvements and bring under one roof, the jobs of planning, developing, implementing and managing all public and private transport infrastructure and policies. The 67-kilometre MRT line will be extended to 160-kilometre MRT and LRT lines by the year 2000, excluding the possibility of possible LRT connections to the HDB new towns.

Clearly, government policies in transportation are tied up with urban reform and the latter is mainly concentrated in the CBD. There has been a trend towards decentralisation of commercial activities away from the CBD in the 1980s, in response to both road congestion and escalating rental costs. The latest move by the URA is the planning of four new regional centres to encourage businesses to decentralise and hence, to make the most of infrastructure like the MRT and expressways. Radiating from the central region are four centres to serve about 500 000 people (see Chapter 2). The concept thus has a dual purpose of diversifying activities which would lessen traffic congestion and cutting down urban commuting to a minimum. Government offices are also relocating to outside the CBD partly in view of this need to decentralise and also to reap higher economic returns by renting out existing government properties within the CBD.

Land for air and sea port development also competes with other land uses. Singapore derives a large part of its comparative advantage and attraction from its infrastructure in communications. Being one of the premier airports and sea ports by international standards, there can be no let up on the maintenance standards of such infrastructure. Again, without a proper market to set land price, any cost-benefit exercise to determine or rank opportunity costs is futile. Increasingly, the problem is becoming a real one as statutory boards like the Port of Singapore Authority (PSA), the Civil Aviation Authority of Singapore (CAAS), the Mass Transit Railway Corporation (MTRC) and Singapore Telecom make their massive land claims for expansion. Apart from evaluating economic costs and benefits, social costs and benefits increasingly enter the picture as social problems, rather than economic ones, become more dominant in Singapore in the next decade.

Land for Agriculture

Land for agricultural use has been deliberately restrained on the premise of high opportunity costs. Moreover, an agricultural activity such as pig rearing, is pollution-prone. Only high-technology agrofarming such as hydroponic cultivation, orchid and aquarium fish breeding are encouraged. The problem of food security is paid little due as Singapore practises a free trade policy to source from all over the world and it probably believes that there is minimal threat to food supply.

Extension of Land Resources

While land is a finite commodity despite technological marvels, there is scope in widening the natural confines of territorial space. The growth triangle involving Singapore, Batam and Bintan Islands and others in the Riau Province in Indonesia and Johor in Malaysia marks the beginning of a more extended case of a regionalisation policy launched in 1993 (Low et al. 1993). Apart from the option of Singapore decanting its labour- and land-intensive activities to other growth nodes to prevent overheating of the economy, there are further dividends to be reaped from such efforts at economic cooperation. Singapore's attractions combined with those of Riau Province and Johor make it many times more attractive in a region that is widely recognised as one of the most dynamic growth areas. Land and related resources feature significantly as motivations from the Singapore side.

The growth triangle and further regionalisation of GLCs, multinational corporations (MNCs) and local companies into China, Indochina and the Indian subcontinent constitute an interesting experiment of transborder production. This enables Singapore to "borrow economic space" and at the same time, accelerate the learning curves in the industrialisation efforts of these countries. The concept has been very much initiated by the private sector pushing out of Singapore by high factor costs and inadequate supplies. The government's encouragement and official blessing of the growth triangle and regionalisation policy have accelerated the pace of regionalisation, as well as provided focus and direction.

The transformation of the physical environment in Singapore as both a cause and effect of its socio-economic development, is moving towards a phase emphasising more quality and gracious living. The "Garden City" concept initiated by the then Prime Minister and now Senior Minister, Mr Lee Kuan Yew in 1963, is progressing towards becoming the first developed city on the equator. This aims for a population of four million people to enjoy bigger and better quality homes and the better relocation and organisation of industries. New land reclamation which would increase land stock by 2.7 per cent or 9100 hectares, is a major strategy. This switch from quantity to quality in physical planning is in response to greater wealth and affluence accumulated with economic growth. In making this switch, innovative and bold ideas require participation and infusion from the private sector.

Institutionally, there has been a change commensurating with this emphasis. In 1989, two bills were passed in Parliament to redefine the scope of planning functions in the Planning Act and the URA Act, paving the way for all planning to be under the URA, amalgamating it with the Ministry of National Development's Planning Department and its Research and Statistics Unit. This consolidation brings together the experience of the URA in urban development, the Planning Department's in strategic planning, supported by data collection and analysis of the Research and Statistics Unit. The URA is also working more closely with the private sector to come up with a more imaginative land use package to suit the economic and social needs of Singapore in the year 2000 and beyond.

Conclusion

Being a city-state, Singapore is unique, with only Hong Kong as the closest parallel. Given Singapore's economic and political circumstances, its built environment is not truly reflective of a free market. For one, land cannot be priced or market determined even though the government is trying to be more realistic about land costs when giving state land to the HDB for its housing programmes. Such land use and housing policies are more reflective of socialist principles than capitalistic market economies. These are further reinforced by very restrictive rules in the 1960s and 1970s regarding the use of CPF savings for housing, subsidisation based on income ceilings and other eligibility rules and resale rules.

Considerable price distortions and distortions of housing tenure, bundle and location thus resulted as the political goal of 100 per cent home ownership also entered the economics of public housing. The government attempted to correct the artificial demand for public housing only in 1979 and 1981 by liberalising CPF savings to include financing for private housing as well as large increases in prices of public housing.

The future prospects for government intervention in the built environment would probably continue to be strong. There would be continual re-ploughing of leased land back to state such that even if the government does not acquire more land, its land stock would not diminish in the long term. But so long as private sector initiatives and investments in land development are not stifled or discouraged, this dominant state ownership of land is not a fundamental problem. Optimum land use is not merely getting more utility out of land space but also having to plan carefully to integrate island-wide resources. Since the state owns such a sizeable amount of land, this implies private-public sector collaboration. The current mood to privatise and cooperate with the private sector is favourable to such a development. The co-operation ranges from conservation to redevelopment.

This paper has also noted some adverse effects arising from the government's preoccupation with 100 per cent home ownership scheme which may need further intervention and fine tuning. The state will clearly not attenuate its control of land as it is seen as one resource which not only provides the government with overall control of economic and social activities but also a source of financial resources. In an urban city-state setting with no other natural resources, the government is not apologetic about earning economic rents from land and land appreciation due to its infrastructural investment.

The adverse results of the emphasis on land and ownership comes mainly from the speculation by buyers and the resulting asset inflation which is already potentially high in a small urban city-state. The appetite of Singaporeans for real estate and properties have also spilled over to neighbouring countries in Malaysia and Indonesia as well as China.

Regarding the constraint on land space, the horizon is bright with economic prospects offered by the growth triangle encompassing the Riau Province and Johor. The extension is taken a notch higher in the effort to build a second wing through the regionalisation policy. Such prospects have been reinforced formally at the government level. Consequently, Singapore appears to have an "infinite" supply of land when such options of harnessing regional resources on a collaborative basis are possible. In reality, this is conditional on the political dimensions of maintaining cordial relations with the neighbouring countries which is a further test of the political economy approach to Singapore's survival. But no matter how harmonious political relations are, note must be taken of factors of nationalism and the interests of other countries in catching up on development status so as to be able to compete with Singapore's growth path.

In other words, the growth triangle and regionalisation policy should be viewed more practically as only a "window of opportunity" which can be closed either by economic nationalism or economic development in the neighbouring countries. It would be in Singapore's interest to invest both resources and time to take advantage of the "window of opportunity" and to "lock" itself into the development strategies of nearby countries to make itself indispensable as a partner rather than to merely exploit their resources. If Singapore can be proven as an invaluable business partner and source of technology and expertise which can be profitably shared by all, the durability and stability of the growth triangle would be enhanced. Whether domestically or externally, the harnessing of land resources is as much an economic as a political exercise.

ENDNOTES

1 See statistics from Motha (1981), p. 105 and (1982), p. 13 and Lim C.Y. et al. (1988), p. 101.

2 The corresponding percentages for total population housed in HDB flats are 37 per cent in 1970, 73 per cent in 1980 and 89 per cent in 1993. The Homeownership Scheme (HOS) for purchasing HDB flats using enforced savings from the Central Provident Fund (CPF) was started in 1968. The use of CPF savings for housing was liberalised to middle-income Housing and Urban Development Company (HUDC) flats and to private residential

properties in 1981; see Lim C.Y. et al. (1986). The demise of the HUDC is noted in footnote 8.

3 See Ricquier (1985), pp. 17–18. Legislations which affect urban development include the Planning Act, Land Acquisition Act and Control of Rent Act.

4 See Khublall (1984), p. 27.

5 Some studies using the political economy framework include Lee (1982), Pang and Lim (1982), Pang and Chan (1986), Low (1987), Rodan (1989) and Pugh (1989).

6 In the 1984 general election, for instance, an opposition Member of Parliament won a seat in Anson constituency where 80 per cent of the 700 families comprising 1700 voters who were relocated to make way for a container port, cast their "protest" votes against the PAP.

7 The author is grateful to an anonymous referee for this information.

8 As at 1 May 1982, the HDB became the national housing authority when JTC low-cost flats and the Housing and Urban Development Company (HUDC) became defunct because the private sector prices for condominiums and other flats became equal to what the HUDC could produce. All properties under the HUDC were transferred to the Urban Development Management Company (UDMC) for management.

REFERENCES

Briffett, C. (ed.) (1990) *Master Plan for the Conservation of Nature* (Singapore: Malayan Nature Society).

Chan, H.C. (1989) "Politics in an Administrative State: Where has the Politics Gone?", Occasional Paper No. 11, Department of Political Science, National University of Singapore, January, (Singapore).

Chan, H.C. (1989) "The PAP and the Structuring of the Political System" in K.S. Sandhu and P. Wheatley (eds.), *Management of Success — The Moulding of Modern Singapore* (Singapore: Institute of Southeast Asian Studies), pp. 70–89.

Cheah, C.W. (1989) "A Benefit-Cost Evaluation of Public Housing in Singapore" mimeo, Department of Economics, University of Queensland, Australia, January.

Chin, A.T.H. and Koh, L.J. (1989) "Factors Affecting Car Ownership in Singapore", Econometrics Studies Unit Working Paper No. 3/89, Department of Economics and Statistics, National University of Singapore, (Singapore).

Department of Statistics, Singapore (various years) *Yearbook of Statistics, Singapore.*

Department of Statistics, Singapore (1991) *Census of Population 1990, Households and Housing.*

Gamer, R.E. (1972) *The Politics of Urban Development in Singapore* (Ithaca: Cornell University Press).

Housing Development Board (1993/94) *Annual Report.*

Khublall, N. (1984) *Law of Compulsory Purchase and Compensation* (Singapore: Butterworth & Co).

Langford, J.W. and Brownsey, K.L. (1990) *Economic Policy-Making in the Asia-Pacific Region* (Halifax: Institute for Research on Public Policy).

Lee, S.A. (1982) "Singapore's Political Economy — A Corporate Perspective", *Singapore Business Yearbook 1982* (Singapore: Singapore Press Holdings), pp. 35–45.

Leung, Y.K. (1987) *Development Land Development Charge in Singapore* (Singapore: Butterworth).

Lim, C.Y. et al. (1986) "Report of the Central Provident Fund Study Group", *Singapore Economic Review*, Special Issue, Vol. 31, No. 1, April.

Lim, C.Y. et al. (1988) *Policy Options* (Singapore: McGraw-Hill).

Low, L. (1987) "The Economics and Politics of Public Housing in Singapore", *Kajian Ekonomi Malaysia*, Vol. 24, No. 1, June, pp. 67–83.

Low, L. (1990) "Changing the Economic Policy-Making Process in Singapore — Promise and Problems", in J.W. Langford and K.L. Brownsey (eds.) *Economic Policy-Making in the Asia-Pacific Region* (Halifax: Institute for Research on Public Policy), pp. 299–319.

Low, L., Toh, M.H., Tan, K.Y., Soon, T.W. and Hughes, H. (1993) *Challenge and Response — Thirty Years of the Economic Development Board* (Singapore: Times Academic Press).

Low, L. (1994) "A Market System with Singapore Characteristics", Centre for Advanced Studies Source Materials, National University of Singapore, November.

Low, L. (1994) "The Privatization of Singapore Telecom", *Asia Pacific Telecommunity Journal*, Vol. 6, No. 3, July.

Low, L. (1995) "A Review of Car Policy in Singapore", *Wheel Extended*, Dynawood Incorporated for Toyota Company, May.

Ministry of Trade and Industry (1986) *Report of the Economic Committee* (Singapore).

Motha, P. (1981) *Property in the 1980s* (Singapore: Select Books).

Motha, P. (2nd ed.) (1982) *Singapore Real Property Guide* (Singapore: Quins Pte Ltd).

Ooi, G.L. (1990) "Town Councils in Singapore — Self-

Determination for Public Housing Estates", Occasional Paper No. 4, Times Academic Press for Institute of Policy Studies, (Singapore).

Pang, E.F. and Chan, H.C. (1986) "The Political Economy of Development in Singapore, 1959–1985", Paper presented for Institute of Southeast Asian Studies Workshop on Political Economy of Development in ASEAN, 4–5 April, Singapore.

Pang, E.F. and Lim, L. (1982) "Political Economy of a City State", *Singapore Business Yearbook 1982*, Singapore, pp. 7–33.

Phang, S.Y. (1989) "The Effects of Government Policies on Housing, Location and Commuter Decisions in Singapore — Theory and Evidence", Unpublished PhD thesis, Economics Department, Harvard University, Cambridge, Massachusetts.

Pugh, C. (1989) "The Political Economy of Public Housing", in K.S. Sandhu and P. Wheatley (eds.) *Management of Success — The Moulding of Modern Singapore* (Singapore: Institute of Southeast Asian Studies), pp. 833–859.

Ricquier, W.J.M. (1985) *Land Law* (Singapore: Malayan Law Review).

Rodan, G. (1989) *Political Economy of Singapore's Industrialization — National State and International Capital* (London: Macmillian).

Sandhu, K.S. and Wheatley, P. (eds.) (1989) *Management of Success — The Moulding of Modern Singapore* (Singapore: Institute of Southeast Asian Studies).

Tan, A.H.H. and Phang, S.Y. (1991) "The Singapore Experience in Public Housing", Centre for Advanced Studies Occasional Paper No. 9, National University of Singapore, (Singapore: Times Academic Press).

Toh, M.H. and Low, L. (1988) "Economic Planning and Economic Policy Making in Singapore", *Economic Bulletin for Asia and the Pacific*, Bangkok, United Nations, Vol. 39, No. 1, June, pp. 22–32.

Urban Redevelopment Authority (1993/94) *Annual Report.*

Wong, A.K. and Yeh, S.H.K. (eds.) (1985) *Housing a Nation — 25 Years of Public Housing in Singapore,* Maruzen Asia for Housing Development Board, (Singapore).

Yeh, S.H.K. (1984) *Households and Housing* Census Monograph No. 4, Department of Statistics, November, Singapore.

6

Planning of the Built Environment and Business

Ooi Giok Ling

Introduction

In the South-east Asian region, where city infrastructure lags behind investment and where it is obvious that there are high operating costs and social costs such as those incurred because of severe urban congestion, it is not surprising that privatisation and other such policy options have figured frequently in infrastructural development plans to support the boom that the regions' economies are experiencing.[1] Furthermore, "overburdened central ministries often provide services and infrastructure inefficiently".[2] Moreover, governments face increasing financial constraints in meeting public service demands, which is the result of previous nationalisation policies.[3]

In noting that "throughout the history of the United States, for example, private organisations have been heavily involved in building toll roads, turnpikes, canals, waterways, railroads, and airports", there is recognition that such private-sector participation in South-east Asia has assumed a variety of different forms such as corporatising and marketising public agencies to require them to cover costs and operate more efficiently.[4] At the same time, there are urban services that the private sector might not be in a position to finance including huge infrastructural projects like mass rapid transit systems. There are also services such as pollution or disease control which do not promise profits.

Among the other developing countries' current re-assessment of the private sector's role in infrastructural development, Singapore does stand out as one city where public sector provision of almost all urban services and infrastructure has actually been successful and it has been observed that the economy would simply not exist without the role played by the state in the development of the urban facilities — transport, industrial estates, airport, port, public housing that are already in place.[5] Yet the ruling PAP's approach has always been to delegate some of the responsibility for social programmes such as housing and health care, to employers and workers themselves.[6] The government has, therefore, not only provided for public needs for urban services and collective goods but also those of the business sector's. This paper intends to examine the implications of the central and dominant role played by the state in meeting businesses' needs.

The following discussion focuses on the importance which has been attached to accommodating the needs of the business sector in the planning and development of Singapore's built environment. Some of the concerns have already been dealt with in the other chapters of the book where the dominant role of the state and hence, the public sector in the planning of the built environment in Singapore has been discussed. The prominent role played by the state arises as a matter for careful study and analysis not only because the public sector is the major land-owner in the city-state but also the major force in land-use and other physical planning and development decisions.

The challenge in this study is to answer some basic questions from the business point of view — how the planning and development of the built environment has accommodated business needs in the city-state of Singapore as well as whether such effort has kept pace with changing business conditions and hence, expectations, not only locally but also globally. In the following discussion, a review of a number of international polls which have been carried out shows that Singapore is ranked relatively high as a country and city for doing business.

The concern for the business sector was evident early in the process of planning for Singapore's physical transformation.

This process of transforming Singapore from former British colony and one of the Empire's Straits Settlements to modern nation-state has been bound up with the aim of encouraging foreign investments and promoting the industrialisation drive initiated early in the 1960s. Lacking as it did, a natural resource base or large market to offer the businesses that Singapore was striving to attract, it was inevitable that effort was then concentrated on planning for a built environment to accommodate the business sector's needs.

In discussing the city-state's built environment of the future, the bureaucrats and leaders have provided several goals which have implications for the business sector. "Singapore seeks to be a hub city for the region and the world in a growing interdependent global economy. ... We will play a catalytic role in integrating the region together and servicing them and helping them to grow," was the recent announcement by Brigadier-General (NS) Lee Hsien Loong, Deputy Prime Minister and the then Minister for Trade and Industry.[7] The goal of becoming a hub city for the region and the world sets the city-state firmly within the global context and the standards against which its physical assets and attributes for business are to be measured. Such a goal and the others which have been explained by the former Chief Executive Officer and Chief Planner of the Urban Redevelopment Authority (URA), Mr Liu Thai Ker, imply changes in the shaping of the built environment which would concern the business community.

Planning for Changing Business Needs

Mr Liu Thai Ker had observed that:

> We are now talking about the second wave of urban improvement which will be going for quality, variety, excellence in architecture and urban spaces. We should also make better use of our natural assets. ... We should also think of ways to bring greater variety to our environment. For example, within the HDB new towns, we are planning to have private developments. We are also introducing business parks because that is something investors want in a better, upgraded industrial environment.[8]

The aims of providing for businesses in the foreseeable future emphasise variety, the involvement of the private sector in public sector developments such as public housing estates and the provision for alternative locations for business in business parks. Business parks are being developed by the Jurong Town Corporation (JTC) to accommodate the needs of modern businesses and to enhance Singapore's position as a total business centre. This business-park concept emphasises a high quality working environment which includes landscaped grounds, high quality and adaptable buildings with modern telecommunication facilities.

The first business park called the International Business Park (IBP) is being developed and comprises a ten-storey business building with a net rentable space of 30 000 square metres. Companies can opt to build their own premises on a number of prepared land sites available.[9] With the introduction of such business parks, the JTC has come some distance since the development of its first industrial estate. The JTC now manages some 30 industrial estates which accommodates some 6500 companies employing a total of 334 000 workers.

A second phase of the Singapore Science Park has started. The park which is located next to the National University of Singapore, is aimed at accommodating research and development (R & D) companies. The aims that have been set down by the Urban Redevelopment Authority (URA), the foremost local authority on land-use policy, as well as the other state agencies such as the JTC and the Economic Development Board (EDB), actually provide an expedient framework within which the relationship between the planning of the built environment in Singapore and its business sector can be examined.

While not an exhaustive attempt, the discussion in this paper will assess the impact that state planning has had in accommodating business needs and the future direction of such planning effort. The discussion, therefore, focuses on the opportunities for the private sector's participation in physical development and property ownership; options in locations and premises which are available for business that would involve considerations of transport and accessibility for the firms and their employees.

The relationship between the business sector and the built environment has numerous ramifications. Its concerns with the planning and development of the built environment are as numerous and varied. Nevertheless, it is possible to distil the major business concerns. Some of these concerns are exemplified by the considerations of the businesses which are contemplating the potential of Eastern Europe as their next stop. Topping the list of difficulties potential investors are being warned of in European countries like East Germany is the issue of property rights. This is followed by further warnings that businesses might have no infrastructure supporting them: no phones, no adequate sewage in some cases, electricity supply problems, trained staff and so on.[10]

If the above are the basic concerns of businesses, Singapore has made ample provision for such needs and is now embarking on what appears to be a new round of city-building to ensure that the future state of the built environment meets changing business needs. The challenge lies in identifying these needs now that the basic concerns have been addressed. There are parallels in public housing in that, having met basic housing needs, the demand has grown for qualitative improvements in such housing and housing environment.

While there is generally a recognition of the city-state's phenomenal success with housing its people and laying down the physical infrastructure in support of its equally successful and dynamic economy in the last three decades, there has been, more recently, a claim for the need to go beyond basic provision towards qualitative rather than quantitative goals in urban design.[11] During the early years in the 1960s, there was recognisably a need for speed in delivery and an emphasis on quantity in the city-state's physical development and the priority it was given over quality. In tandem with the need to provide speedily for homes as well as the needs of industry and business, it was expedient for the public sector to assume a major role.

This paper intends to focus on a number of aspects of the built environment which has implications for the private business sector — the role which the private sector has played in the development of the built environment and relatedly, the impact of the planned urban environment on locational options available to business and the range of options for the development of business premises. This discussion centres on two themes. The first is the need for private-sector participation and the second, the impact of planning policies and practices on business options in the two aspects mentioned earlier—namely, locations and business premises. Before discussing these two areas of business-related development, the paper will briefly review the impact of urban planning practices on the spatial distribution of business in the city-state. Finally in the discussion, matters on the built environment which are of future concern to business will be highlighted.

A series of questions are being asked in this paper. Has business been locationally "strait-jacketed" because of urban planning practices in the last three decades? What has been the impact of these planning practices on business? With the public sector a major player in the shaping of the built environment, has the private sector been able to realise its needs in the range of business premises available? In the bid by the public sector to work more closely with the private sector in the physical development of the city, what are the areas in which they can best cooperate to achieve an urban environment that can better accommodate business interests not only in the near future but right through to the next century.

The Roles of Private and Public Sectors

The URA's sale of sites guided almost all of the commercial development in the Central Business District (CBD) and the Orchard Road Belt and the public sector was the catalyst for the early push into property development, with the Housing and Development Board (HDB) starting on public housing in the 1960s, the Port of Singapore Authority (PSA) developing industrial and warehouse space and Jurong Town Corporation getting into building factories.[12] The observation has been made that, in the last three decades, the public sector architect has been handling seven times the volume of design work in terms of floor area compared to the private sector architect.[13]

The dominant role of the public sector has to be studied in the light of the effects on options for the business sector in terms of considerations such as business locations and premises. Public land-use development planning policy has literally decided where business should go and how much area it should be allocated in the last three decades. While it is true that the more *laissez-faire* approach to land-use development planning adopted by the British colonial adminstration resulted in severe urban congestion and attendant problems largely because of its ad hoc nature, the business sector would probably welcome the city planners' intention of introducing more flexibility and a wider range of options than have been available in the last three decades of post-Independence planning.

Consequently, the translation of plans for developing Singapore into a tropical city of excellence or a global city has implications for the business sector. If the rationale of the plan is to promote economic growth or gracious living or both, then there would still be implications arising from exactly what is entailed in making the city tropical and global as its planners embark on the next phase of physical development. The aim of being excellent and that of introducing variety into the built environment prompt the question of whether the public sector will retain its current major role in urban planning and development or whether more participants from the private sector, especially local firms, will be involved. A related question would be the future role of foreign architects and designers who have been responsible for the design of some of the city-state's major buildings.

There is a trend towards the involvement of the private-sector designers in public sector projects. The process of drawing up development guide plans for three new development areas in the city-state involves private-sector designers. Dialogue sessions have been held between the private and public sectors to review building controls.

Similarly, the aim of involving the private sector in public-sector development projects imply a cooperation premised on a two-way exchange between the sectors, a greater cross-fertilisation of ideas and a more open as well as discursive planning and urban design culture than has otherwise been the case in the past. With the proposal to provide businesses with improved environments, urban planners might well have to consider how best they can accommodate the ever-changing needs of businesses if there were to be no greater cooperation between public and business sectors in the future.

In his discussion on the shape of things to come, Mr Liu Thai Ker has indicated that:

> The second big wave of development will be in the near future. The city centre is getting so big, property values are getting high and congestion has become a problem. We are now seriously thinking of decentralisation in a sense that, wherever possible, we would like to locate some businesses outside the city. By doing this, we would like to give them an alternative. ... The way business is done today, service- and marketing-oriented businesses will be in the central area. We want to encourage back-room type businesses to move to the regional centres.[14]

Five such regional centres are in the offing. The island, as the writers in the earlier chapters have already discussed (see Chapters 1, 2 & 5), has been divided into five regions — the central and four other regions outside. Each is planned to have a population of half a million in order to make the region an independent city. Within each region will be a centre served by major roadways and the Mass Rapid Transit (MRT) lines. The intention, according to Mr Liu Thai Ker, is to provide a wider range of alternative locations for businesses.[15]

The intentions of the public planning sector, might well augur for the business community more of the kind of options arising from policies like the Area Licensing Scheme (ALS) in the Central Business District (CBD) (see Chapter 4 for the discussion of the policy). In the aftermath of the implementation of the policy, property consultants saw a big relocation of companies from the CBD to the CBD fringe.[16] This impact reportedly lasted until the introduction of other policies on land transport, including the recent changes to ALS charges.

State policies, therefore, wield substantial impact on locational options among businesses. In the city-state of Singapore, the massive spatial re-organisation exercise initiated in the 1960s has had immense impact on where industries and other business firms are located. The success with the re-distribution of commercial and industrial land-uses in tandem with the city's urban re-development programme appears to be reflected in the city-state's international ranking as a business centre.

World City: The City State in a Global Context

Global comparisons of cities have recently produced a spate of lists of the world's best and worst cities in terms of such criteria as livability and costs.[17] The lists are drawn up for the benefit of a world and particularly, a business community which is increasingly mobile and taking to world travel on an unprecedented scale. It is particularly the case with the business community because of the trend towards globalisation through trade and investment across national borders.

Among the EDB's strategies for the year 2000 are Manufacturing 2000 or M2000 which aims to sustain the manufacturing sector's contribution to GDP at more than 25 per cent by focusing on upgrading and restructuring the city-state's manufacturing clusters in order to encourage higher value-added activities and employment.[18] Another major developmental thrust for the agency which spearheaded Singapore's industrialisation drive in 1960 is International Business Hub 2000 or IBH2000 aimed at promoting Singapore as a regional gateway for the service sector.

Singapore's physical planners realise that there is more to the bid to joining the ranks of the world's great cities than providing a world-class airport and a technologically advanced efficient telecommunications network.[19] In its latest survey covering some 50 countries, the US-based Business Environment Risk Intelligence (Beri) placed the city-state as one of the three top business centres, ranking Singapore only after Switzerland and Japan and on par with the united Germanys.[20]

In yet another survey of the most liveable cities in the world conducted over a two-year period by the Washington-based Population Crisis Committee among 45 countries, Singapore scored 79 out of a possible 100 points. The world's 100 largest cities were included in the survey and ranked according to their scores. The evaluation was based on the so-called culturally neutral criteria — murder rate or public safety, health using infant mortality rate, education and housing (space per person and utilities at hand), cost of food, telephone services, communications, traffic, air quality and noise levels.[21] By the measure used in this survey, Singapore was placed in the category of cities with very good standards of living where there were four categories ranging from very good to poor standard of living. These surveys provide a spectrum of the criteria which have been used to analyse business conditions as well as cities, or more specifically, the urban built environment. Judgement of the city-state as a business centre was based on politico-legal and economic criteria whereas the assessment of its liveability stressed physical infrastructure and environmental conditions.

In identifying the four stages in the physical development of Singapore, a local architect, Tay, has also singled out the major political and economic forces which have influenced the periods which he has named Malayanisation, 1955–65; the evolution of the independent island city-state and industrialisation of the economy, 1965–70; the global-city concept, international corporatisation and development as a financial centre, deepening of corporatisation and the growth of "high-tech" industry, 1980–85.[22] From another perspective, a study of the world's capital cities will highlight their many varying features and claims to global status.

The major European capitals such as Paris, Rome and Vienna can, arguably, lay claim to being European Renaissance cities with a grandeur that only a long history and long-established cultural heritage can imbue. New York and Seoul have been considered the Western and Eastern versions of industrial megapolises. Hong Kong and Taipei are cities reflecting the single-minded pursuit of maximising the utility of urban space. Vancouver, on the other hand, appears to have stressed gracious living and a city that has capitalised on its natural assets of coastline location and scenic physical terrain. Setting aside the problems of traffic congestion, crime and air pollution, these are cities which could claim to have achieved excellence and a global-city image in their own way within the constraints that each faced. Other cities claim to be strategic banking centres — Frankfurt, New York, Singapore, Hongkong and Tokyo or strategic nodes in port facilities — Rotterdam, Hamburg, Singapore, Taiwan and a few seaboard cities in America.[23] The corollary is that a city's claims to global status must generally rest on a package of "competitive" business functions as well as its living standards.

Now that the physical infrastructure is in place in Singapore and the city has gained a reputation for efficiency in the delivery of urban services, it seems apposite that the private business sector be given the opportunity to play a greater role in the physical development of the city. This need tallies well with the public sector goals of introducing quality to the development of the built environment with emphasis on variety and improvements for business environments.

Even while recognising that Singapore's achievements in urban growth and the development of the built environment have been phenomenal in the last three decades, it is the business community's contention that there should now be more consideration of the adjustments that must be made to enable the city to scale the height of excellence and achieve global status. It is apt that, with an emphasis on further economic growth through globalisation, there is a similar pursuit by the city's planners to become global as well. After all, the city's physical development in the last three decades has been an integral part of the country's growth from entrepot port to a newly-industrialising economy.[24]

The locational and other options which the physical planning bureaucracy would like to provide the business community in Singapore is one of the major themes and concerns of this paper. Major aspects of the landownership pattern in Singapore have already been discussed (see Chapter 5). In that paper, Low also discussed the growing proportion of land in the ownership of government statutory boards and the impact on the built environment. Consequently, there followed a minimising of the private sector's role in land-use planning and the shaping of the built environment.

The Impact of Land-use Policies and Planning Practices

Business in this paper is defined widely. Encompassing private-sector developers and designers, it also includes all commercial activities such as retailing and industry. In the 1982 Building and Land-use Survey Report, commercial activities were defined as including offices, shops, markets and hawker centres, restaurants, entertainment and recreation centres and hotels. The definition of industrial land-use included flatted factories (of four storeys and above), row-terrace factories, detached and semi-detached factories as well as temporary industrial buildings and buildings such as shophouses used for industrial purposes. However, the discussion will, in the two main sections of this paper, focus on specific segments of the business sector such as retailing and the hotels.

The concentration of business and the better part of the population in the Central Area during the 1950s and early 1960s has already been discussed (see Chapter 2). Consequences arising from the indiscriminatory mix of shops, homes and industries in the Central Area have encouraged the expert groups working on the Master Concept Plan to advise a massive spatial redistribution exercise. In this exercise which has spanned the last three decades, a new spatial ordering has emerged. Land and locations for business of the residential population have been allocated according to plan. The new spatial order, therefore, sets down specific locations which are to be used for business. There are also specifications for the development of business premises. In the Master Plan, the plot ratio and density of developments are prescribed. According to one observation, regulatory mechanisms such as these have proven to be useful in staying with the worse results of the "boom and bust" cycle in the property market.[25] The Singaporean and Malaysian governments have been found to have taken urban zoning seriously compared to other Asian cities. The two governments have also reportedly been more vigilant over banks' real-estate portfolios, sounding the alarm on over-development a good deal sooner than regulators elsewhere.[26] At the time the Master Plan was being drawn up in 1955, it was observed that:

> The main planning consideration in relation to business users is the volume of traffic they create, its effect on the capacity of roads giving access to the Central Areas and on car parking facilities. So important is commercial activity in Singapore that we are reluctant to impose any control which might limit its traditional freedom. Nevertheless, an acute traffic problem exists and is growing still more acute.[27]

The traditional freedom which business had been observed to enjoy originated from Sir Stamford Raffles' land-use policy which simply decreed that "... in allocating lands, first preference should be given to merchants, second to artisans and third to farmers ...".[28] Apart from the early emphasis on the needs of merchants and hence, business, this approach to land-use development had engendered the practice of developing land to meet requirements as they arose.[29] Such an approach had resulted in major problems for land-use developments by the late 1950s.

The results from the implementation of the Master Concept Plan and its quinquennial revisions has created a new spatial ordering which is best illustrated in the following comparison of Central Area land-use distribution between 1967 when an earlier building and land-use survey was done and 1982 (see Table 6.1). There has been some ten per cent reduction in the land developed for commercial use within the Central Area between 1967 and 1982. This has been attributed to the demolition of old shophouses in the 1982 report on the land and building use survey. In terms of floor area, however, there has been a tremendous leap in growth of commercial space. Of a total of 8.8 million square metres of commercial floor space, 53 per cent or 4.7 million square metres are located within the Central Area. The 1982 land and building use survey report rightly observed that this re-emphasised the importance of the Central Area as the commercial hub of the city-state. With only 18 per cent of the land in the city-state, the Central Area accommodated 53 per cent of the commercial floor space. This included 71 per cent of the total office floor space, 56 per cent of the entire city-state's hotel floor space or 78 per cent of the total number of hotel rooms, 45 per cent of all shopping floor space and 37 per cent of all restaurant floor space.

In terms of spatial distribution, industries have also been allocated specific locations mainly outside of the Central Area. The largest industrial estate is the Jurong Industrial Estate spanning 1214 hectares in the western end of the main island. Another major industrial area is the Kallang Basin in the east. There are also other industrial estates which are intermediate in size and still others that are small. These estates are mostly located in areas outside of the Central Area. Many are sited in and near to public housing estates.

Table 6.1 Land-use Distribution in the Central Area, 1967 – 1982

Land-use Distribution	**1967 (%)**	**1982 (%)**
Residential	10.8	10.3
Industry	3.6	1.0
Commercial	24.9	15.1
Transport Utilities/	22.8	21.6
Communication	1.2	0.9
Institutions	11.5	6.1
Military	1.7	—
Recreation	8.5	6.8
Cemeteries	0.5	0.2
Vacant buildings & buildings under construction	3.9	6.2
Vacant land & other undeveloped land	10.6	24.8
Warehousing	—	1.4
Special Use	—	0.4
Water Areas	—	1.9
Total percentage	100.0	100.0
Total hectares	724.7	1020.0

Source: *Land and Building Use: Report of Survey, Singapore, 1968 and 1982 Land and Building Use, Report of Survey, Singapore.*

The premises provided to house industry have already been listed above. It is significant that this range of premises also officially defines industrial land-use. Industry has, therefore, been allocated location and premises within the planned spatial order. Such planning appears to allow relatively little room for manoeuvring. Certainly, there are no backyard addressess for industries to the extent which is characteristic of say, Hong Kong or Taiwan.

Commercial land-use which has already been defined, has spread from its earlier concentration in the Central Area. Between 1967 and 1982, there was a remarkable increase in the floor area under commercial use. In the Central Area, this increase has been achieved in spite of a decrease in the land area occupied by commerce. Together the four main commercial activities — offices, shops, hotels and restaurants dominate the floor space within the Central Area. The phenomenal increase in floor space in use by each of these activities is illustrated in Table 6.2.

Table 6.2 Commercial Floor Space in the Central Area, 1967–82

Commercial Use	**Increase in Floor Space (1967–1982) (%)**
Offices	340
Shops	48
Hotels & Restaurants	566

Source: *Land and Building Use: Report of Survey, Singapore, 1968 and 1982 Land and Building Use, Report of Survey, Singapore.*

Office developments occupied practically half of the commercial floor space in the Central Area, 46 per cent. Since 1967, this has meant an annual increase in floor space of some 120 266 square metres. Nearly half or 49 per cent of the office space has been developed in what is known as the Golden Shoe area which is where the financial district of the city is located. Similarly, the hotel developments have been clustered along Orchard Road and, to a lesser extent, along Havelock Road.

In terms of spatial distribution, commerce has been encouraged to remain in Central Area locations. Until 1983, when it was standardised to 23 per cent for all properties in the Central Area, commercial premises were given a 12 per cent concession on property taxes. Industry which was being encouraged to shift to industrial estates outside the Central had to pay the full rate of 36 per cent in property taxes.[30] An exception among commercial activities being encouraged to stay in the Central Area would have been the small and marginal retailers whose shops were located in areas that had been slated for redevelopment. Many of these were given resettlement incentives to retire. Yet, residential development in the areas outside of the Central Area created business opportunities for small shopkeepers and family-based enterprises which had not been available earlier. With rent control in force since 1947 in the Central Area, small businesses would have been able to obtain premises earlier only through the payment of "tea" money involving relatively large sums that were prohibitive for most small businesses.

Locational Options for Businesses Big and Small

The discussion in this section concentrates on one question. What is the range of locational options available to businesses? Before focusing on a specific aspect of commerce — retailing — the options available to the commercial sector will be discussed first.

The results of urban planning policies are reflected in the creation of the financial district, the shopping and hotel belt and other areas of specialised land-use within the Central Area. Such an approach to land-use allocation has also meant an emphasis on locational specialisation. Especially for commercial activities like hotels and large-scale retailing, its pattern of spatial distribution within the Central Area is clustered around specific locations such as Orchard Road. Such spatial clustering has its advantages in terms of the supply of services and affiliated facilities like shopping and recreation. This would, in part, explain the "market" tendency to expand and build new hotels in the areas where other hotels are already concentrated.

In the past, the practice has been to encourage commerce to remain within the Central Area although zoning and other development regulations apply at the same time. Recently, however, there has been the tendency for the URA which controls Central Area land-use developments, to discourage hotel expansion and new building in the locations where they have previously concentrated. Such a move reflects the balance which the authorities have to maintain between addressing business needs in terms of timing, space and locations.

Not surprisingly, there has recently been renewed interest in working towards closer cooperation between the private and public sectors in both the planning and development of the built environment. A need for such cooperation was recognised by Low in Chapter 5. Such cooperation has already been noticeable at the level of implementation of policy. Where areas have been pegged for conservation, the URA has allowed the private sector to participate in conservation work. The question, now, is whether such cooperation is likely to be extended to the level of policy-making on the built environment. An associated question would be on the approach towards the resolution of conflicts that are likely to arise between public and private sectors on the matter of reconciling preservation in the built environment with the need for businesses to expand or build anew.

A number of other related questions arises when one studies a specific segment of commerce — namely, retailing. In focusing on small-scale retailing, the range of locational options to which it has access to is essentially, either a shoplot in a shopping centre, a shop in a public housing estate or a privately developed shophouse. Following Sim, there has been a boom in the development of planned shopping centres in the last three decades.[31] There has, similarly, been rapid development of planned shopping facilities in public housing estates where 87 per cent of the Singaporean population live.

A choice of locating their business in public housing estates means that the small shopkeeper can select either a local, neighbourhood or town centre shop. The provision of local shops has been discontinued since 1985. In terms of spatial distribution, 50 per cent are located within the neighbourhood centres, 20 per cent in town centres and 30 per cent are local shops. Within this choice, there is a range of other options regarding shopfront orientation.

A study of small shopkeepers showed that, of the options available in the planned context within which they must select a shop location, they vie for shops oriented towards pedestrian malls, hawker centres and markets all of which are places and facilities that generate high volumes of pedestrian traffic.[32] Less popular are the shops oriented towards main roads and secondary estate roads and playgrounds.

The study shows that operating within the selection available for shop locations, the small shopkeepers have opted for locations which best accommodate their budgets as well as requirements arising from their type of trade and clientele expectations. Spatially, shopkeepers plying various types of trades have distributed themselves among the different types of centres such that the largest proportion of local shops, 20 per cent, retail provision goods. The proportion of provision shops declines in neighbourhood centres where there are only 18 per cent. In the town centres which are served by supermarkets and emporia, the proportion of provision shops was only one per cent. In reverse, shops selling high-order goods like textiles and personal effects (e.g. watches, clothing accessories) increase in proportion. There were only two per cent of local shops retailing these goods compared to nine per cent of the neighbourhood centre shops and 42 per cent in the town centres.

This case-example of small shopkeepers illustrate a series of issues of relevance to business in general. The planned urban environment within which business must make its locational choices offers a finite number of options. To compete, business, generally, and the small shopkeeper, in particular, has to find the shop location with the best combination of rental level and clientele traffic appropriate for the trade being plied. For the small shopkeeper, the combination has to maximise convenience to its target clientele and minimise costs. In short, business and in particular, small business, has to rely on an enlightened planning bureaucracy for latitude in its own organisation.

Large-scale business has the further option of competing and developing in locations of its choice, albeit within the tenets set down by planning policies. Shopping centres and shopping centre-cum-office complexes have been developed within the Central Area and outside by private businesses. In this instance, there are more options in locations as well as type of premises. Such options and the speed with which they can be realised depend again on the relationship between public and private sector. Market interests which emphasise time in terms of costs have to be accommodated as well as the objectives of a planning bureaucracy emphasising conservation or other policies that have been set down in the Master Plan and its quinquennial revisions as well as a number of other concept plans that have emerged since then.

Development of Business Premises

The development of business premises by the private sector is subject to planning and building approval. The normal relationship between the private and public sector *vis-a-vis* the development of business premises would be for the former to satisfy the latter that it conforms with policies on both land-use and design.

The definition of industrial land-use illustrates again, the range of options in trade premises which are available to industries in this planned context. Industrial premises have been standardised into a stock set of options involving either operating in a flatted, terraced, semi-detached or detached factory.

In another example, retail planning for public housing estates follows standards of provision which have resulted in a range of retail premises that are listed in Table 6.3. Given the planned nature of the types of premises available to small businesses in public housing estates, and often the uniformity of the premises in terms of appearance, small shopkeepers can vary their shops' look only through changes in shopfronts or the shopping atmosphere, that is, air-conditioning. There are, however, the HDB rules regarding such changes which could have a deterrent effect on the shopkeepers. These require the shop tenants to restore the fronts to their original condition upon ending of the leases.[33] Further regulations apply to the changing of the facades of the shop premises.

Table 6.3 Selected Planned Retail Premises in Public Housing Estates

Type of Premises	Floor Area (m^2)	Planning Standards (Shop to dwelling units)
Shops	30–400	1 to 70
Kiosks	5–15	1 to 60
Eating Houses	450	1 to 750
Restaurants	90–2000	1 to 1000
Mini-Markets	450	1 to 6000

Source: Tan et al. (1985), p. 103.

Within the Central Area, large-scale businesses which involve retailing, hotel and office developments have the option of developing their own premises. The option is subject to prevailing policies on development of the built environment. While these policies have emphasised zoning, plot ratio and density as prescribed in the Master Plan, there are other emphases which have emerged. Among the new emphases would be public sector norms on aesthetics and the gazetting of areas for conservation with the aim of preserving old buildings which have historical significance and are of value to the country's heritage. The winning of tender bids to develop URA sales of sites, for example, is not premised on the amounts of the bids alone. The URA decides on the bids based on criteria which range beyond the size of the bids and would probably encompass the plan to develop Singapore into a tropical city of excellence. One way of involving the business sector in the city's development would be for the public sector to be more explicit about the criteria adopted in the granting of planning and building approvals. With the increasing emphases on conservation and urban aesthetics and a tradition of options within limits for the business community, it is, perhaps, timely and opportune to involve the private sector in working towards a widening of options for business in Singapore.

Business and the City

Judging from the success with the spatial re-distribution of industries and small-scale commerce previously concentrated in the central area, state agencies responsible for different aspects of business development in the city-state have good reasons to believe that the fundamental provisions have already been addressed and planning can proceed to the next phase of qualitative improvements as well as fine-tuning of the provisions already in place. Urban redevelopment has re-vitalised the city centre which now accommodates a thriving financial business district as well as a hotel-cum-shopping belt.

The removal of small-scale retailers and industries to areas outside of the CBD has had positive spin-offs. Small shopkeepers and medium-sized retail outlets located in neighbourhood and town centres in public housing estates complement larger retailers located in the central area. A relatively large number of the small shopkeepers were small-scale retailers resettled during urban redevelopment.

Industries that were encouraged to relocate to purpose-built premises in planned industrial estates appear to have fared equally well in terms of the numbers of small- and medium-sized firms now distributed in these estates. The approach of distributing the industrial estates in all the public housing areas planned and developed by the HDB has meant an increasing range of locational choices, particularly, for small firms. Now that the HDB has introduced schemes to sell shophouses and even stalls to its former tenants, the constraints arising from the agency's management regulations for these premises would ultimately be removed.

Larger firms, particularly the multinationals, are located mainly in the Jurong Industrial Estate, the largest estate developed and managed by JTC. Due to its location away from major public housing developments, many of the firms have found it necessary to provide their own transport services for staff. Since the MRT lines cover only the fringe of the estate and public bus services do not operate frequently due to the lower population density in the area, firms have discussed a growing concern about transport provision for staff. Considering the bid to plan in the next phase for qualitative improvements to physical environments in demand among business firms, city planners might well have to consider a closer coordination between commercial and industrial land-uses and public transport services.

Conclusion

There is a basis to argue for a widening of options for business to achieve the variety, private sector involvement and a world-class business environment which the public sector has indicated that it is willing to consider providing, in its push towards global-city status. This is especially in the light of the recent trend towards greater cooperation between private and public sectors. In this study, various areas have been highlighted where such options could be considered, in the design of new development areas, locations and the development of types of business premises. It is not an exhaustive list.

To widen the options in both these areas, there has to be greater cooperation between private and public sectors. This would require development policies which accommodate both market and public interests. In this way, it would augur a built environment which is accommodative of business interests as well as the needs of the city's people.

ENDNOTES

1 See Richardson, H.W. (1993), p. 53.
2 See Rondinelli, D.A. and J.D. Kasarda (1993), p. 134.
3 Ibid., p. 138.
4 Ibid., p. 135.
5 Castells, M., Goh, L. and Kwok, R.Y.W. (1990).
6 Lim, L. (1989).
7 See *Business Times*, 7 December 1990, p. 28.
8 In *Building a Nation.*
9 See Ministry of Information and the Arts (1995), pp. 112–113.
10 See *Business Times*, 30 November 1990, p. 11 and 4 December 1990, p. 16.
11 Tay, K.S. (1990).
12 In *Building a Nation*, op. cit., pp. 41–42.
13 Tay, K.S. (1990), op. cit.
14 *Building a Nation*, op. cit., pp. 20–21.
15 Ibid., p. 42.
16 Ibid., p. 21.
17 See *Straits Times*, 5 December 1990, which reported Singapore as liveable city in terms of noise pollution. Singapore did not quite make it to the list of the most expensive cities.
18 See Ministry of Information and the Arts (1995), pp. 106–107.
19 K. Corey (1991).
20 See *Straits Times*, 8 December 1990, p. 47.
21 See *Asiaweek*, 7 December 1990, p. 15 and *Business Times*, 7 December 1990, p. 28.
22 Tay, K.S. (1989), pp. 860–878.
23 Yong, P.A. (1990), op. cit.
24 See *Business Times* May 31 1994, pp. 1 and 30.
25 See Wong, A.K. and Ooi, G.L. (1989) "Spatial Reorganisation", in K.S. Sandhu and P. Wheatley (eds.) op. cit., pp. 788–812.
26 J. Friedland (1990).
27 Ibid.
28 See *Master Plan: Report of Survey (1955)*, Government Printing Office, Singapore, p. 4.
29 Ibid., p. 8.
30 Lim, W.S. and P. Motha (1979).
31 Sim, L.L. (1984).
32 Ooi, G.L. (1986), pp. 369–382.
33 Chong, K.C., Tham, Y.F. and Shium, S.K. (1985).

REFERENCES

Building a Nation — 25 Years of Achievement in Singapore's Building and Construction Industry, a supplement to Building and Construction News, August 1990, p. 19.

Castells, M., Goh, L. and Kwok, R.Y.W. (1990) *The Shek Kip Mei Syndromes — Economic Development and Public Housing in Hong Kong and Singapore* (London: Pion).

Chong, K.C., Tham, Y.F. and Shium, S.K. (1985) "Housing Schemes", in A.K. Wong and S.H.K. Yeh (eds.) *Housing a Nation — 25 Years of Public Housing in Singapore*, Maruzen Asia for Housing Development Board, (Singapore).

Corey, K. (1991) "Information Technology Planning — An Assessment of Singapore Experience", Paper prepared for presentation at the Third Asian Urbanisation Conference, 2–4 January, New Delhi, India.

Friedland, J. (1990) "Bubbles Bursting", *Far Eastern Economic Review*, 15 November, p. 71.

Land and Building Use: Report of Survey, Singapore, 1968 and 1982 Land and Building Use, Report of Survey, Singapore.

Lim L. (1989) "Social Welfare", in K.S. Sandhu and P. Wheatley (eds.) *Management of Success — The Moulding of Modern Singapore* (Singapore: Institute of Southeast Asian Studies), pp. 171–197.

Lim, W.S. and P. Motha (1979) *Land Policy in Singapore* (Singapore: Design Partnership).

Master Plan: Report of Survey (1955) Government Printing Office, (Singapore).

Ministry of Information and the Arts (1995) *Singapore 1995*, (Singapore).

Ooi, G.L. (1986) "Implications of Retail Planning and Development on Small-Scale Retail in New Towns", Paper presented at the 1st International Convention on Urban Planning, Housing and Design, 17–18 July, Singapore, pp. 369–382.

Richardson, H.W. (1993) "Efficiency and Welfare in LDC Mega-Cities", in J.D. Kasarda and A.M. Parnell (eds.) *Third World Cities — Problems, Policies and Prospects* (California: Saga Publications), pp. 32–57.

Rondinelli, D.A. and J.D. Kasarda (1993) "Privatisation of Urban Services and Infrastructure in Developing Countries — An Assessment of Experience", in J.D. Kasarda and A.M. Parnell (eds.) *Third World Cities — Problems, Policies and Prospects* (California: Sage Publications), pp. 134–160.

Sim, L.L. (1984) *A Study of Planned Shopping Centres in Singapore* (Singapore: Singapore University Press).

Tan, T.K.J., Loh, C.T., Tan, S.A., Lau, W.C. and K. Kwok (1985) "Physical Planning and Design", in A.K. Wong and S.H.K. Yeh (eds.) *Housing a Nation — 25 Years of Public Housing in Singapore*, Maruzen Asia for Housing Development Board, (Singapore).

Tay, K.S. (1989) "Architecture of Rapid Transformation", in K.S. Sandhu and P. Wheatley (eds.) *Management of Success — The Moulding of Modern Singapore* (Singapore: Institute of Southeast Asian Studies).

Tay, K.S. (1990) "Private and Public Sector Cooperation in the Physical Development of Singapore", Paper circulated and first presented at the Institute of Policy Studies, 10 November, Singapore.

Wong, A.K. and Ooi, G.L. (1989) "Spatial Reorganisation", in K.S. Sandhu and P. Wheatley (eds.) *Management of Success — The Moulding of Modern Singapore* (Singapore: Institute of Southeast Asian Studies), pp. 860–878.

Yong, P.A. (1990) "Singapore Seeks to be a World Hub City: BG Lee", *Business Times*, 7 December 1990, p. 28.

7

Culture and Identity in the Public Housing Environment

Sharon Siddique

Introduction

How do Singaporeans assess the identity side of the coin stamped "Made in Singapore"? Singaporeans welcome millions of tourists every year and their affluence allows them to travel too. Thus, they have no difficulty identifying each other (and with each other) when faced with non-Singaporeans.

If the homogenising outside reference point is removed, however, Singaporeans are equally aware of their internal diversity. Linguistic complexity, religious variety and ethnic differences are part of their everyday lives.

The perpetuation and support of this diversity also form part of the government policy. Since independence in 1965, there has evolved a linkage between each of the major constituent races — Chinese, Malay and Indian. For the Chinese community — 76.7 per cent of the population in the 1990 census, Mandarin has been designated as the mother tongue; for the Malay community (14.6 per cent) the Malay language and for the Indian community (6.4 per cent) Tamil, the language spoken by the majority within this racial community. A bilingual school system reinforces this relationship between race and mother tongue. English is the first language taught in most schools, with a choice between Mandarin, Malay and Tamil as second language. Learning two languages is mandatory to balance the impact of westernisation through English.

Culture

Much of this internal heterogeneity is linked to larger, more homogeneous external reference points. Malay, Indian and Chinese cultures — whether in dance, art or dress styles are part of the cultural influences by external developments in Malaysia, Indonesia, China, Taiwan, Hong Kong and India. Singaporeans receive three television channels from Malaysia and some homes also receive one channel from Indonesia. This is in addition to the four Singapore channels which regularly screen the latest movies and drama series from Hong Kong, Taiwan, Indonesia and India as well as many Western countries. Singaporeans also have easy access to magazines profiling the latest pop stars from Madras and Taipei as well as performances by visiting Chinese opera artistes and Malay and Indian dance troupes.

Certain general points can be raised pertaining to culture. First, much of the awareness of culture in Singapore is concerned with what might be termed as Culture with a capital "C" and which is related to the arts. This is obviously important, but culture with a small "c" or what might be termed as common-sense culture is a much more general phenomenon. It is learned, in contrast to instinctive behaviour. It is what makes us human — and what creates diversity among the human species. It is about the patterns of thinking, feelings and behaviour which have been socially learned. This culture depicts the pattern of everyday life and the required knowledge in order to live in a society. Common-sense culture is often unnoticed and taken for granted. Its unconscious rules govern not only everyday behaviour but also, at a more profound level, certain pre-conceptions concerning the world.

In sum, Culture with a capital "C" can be taken as one type or, perhaps, more accurately, a concrete expression of common-sense culture. In Singapore, this link seems rather tenuous at the moment. One possible reason could be that the arts are often regarded as traditional, something quite marginal to a modern, technological-oriented society, which can contribute very little to the latter.

This, in turn, is based on the contention that change can be represented along a unilinear, uni-directional continuum segmented into the past, the present and the future and that the traditional-modern dichotomy can be welded to this continuum. It is assumed that what is past is traditional and that the modern is derived from the present. There is also an upward tilt to this continuum, taken for granted, which assumes a constant improvement in the quality of human life. Those in the past were not as well off as those in the present and those who belong to the future will be better off than the present. Therefore, what is modern, it is assumed, must contribute more to the improvement in quality of life. To non-western societies, a logical extension of this is to equate the modern with western, the traditional with indigenous. Consequently, tradition is consigned to the past, or at least marginalised and unfortunately, even by scholars who do research on cultural change.

Environment

There are several examples of how common-sense culture operates in everyday life. These examples will be drawn from what may be termed, the most pervasive Singaporean environment — the Housing and Development Board (HDB) or public housing estates.

Environment, according to Lynch (1980), is the distribution of objects and activities in space and time. Beginning in the 1960s, the HDB has radically altered the living environment of over 80 per cent of Singapore's population. The perennial question is — what is the relationship, if any, between this altered environment and culture and identity?

In a continuum of scholarly opinions on the subject, at one extreme would lie the view that environment is the mirror of culture. In studies on Nias, Bali or rural Thailand, anthropologists read the symbolism behind traditional houses, for example, as that mirroring the cultures of their inhabitants. At the other end of the continuum, more associated with modernity, the physical environment can be seen as an empty vessel which people fill creatively according to the demands of their cultural norms. Some Singaporean architects like to speak of creating a "conducive environment" to allow for cultural development.

Identifying Singapore's place in this continuum is not easy. Certainly, it is not found in either extremes. In addressing the relationship between culture, identity and environment in the Singaporean context, the most that can be accomplished is, perhaps, to present some examples of how the concept of common-sense culture might help to gain new insights into this relationship.

Focusing on the HDB's built environment, three groups may be considered — the designers, that is, the HDB planners and architects, the administrators, that is, the HDB estate/area office staff and now, the Town Council staff, and the residents. Of course, these groups are not mutually exclusive but their tasks and perceptions of the relationship between culture, identity and environment differ.

Needless to say, the three groups also share a great deal in common. Being Singaporeans, they share a complex common-sense culture which could be described as Singaporean. To elaborate on the definition of environment cited earlier, the distribution of objects and activities in space and time needs to be discussed.

Space

Scholarly literature on space, that is, the physical dimension of environment, is littered with terminology — region, area, place and territory (Relph 1976; Lynch 1980). In the HDB environment, there are precincts, neighborhoods, estates and New Towns (Tan et al. 1985). This plethora of terms and rather disparate body of terminology lead us away from formulating fundamental questions regarding space.

To reiterate, common-sense culture is, generally, taken for granted. Cultural orientations generally influence behaviour. Otherwise, people would not be human. Thus, when an architect, for example, picks up a pencil and begins creating the form of a building, his technical skill is still motivated, usually sub-consciously, by the quotidian elements that make him a member of society. When an administrator makes a decision, it is certainly predicated on the practical skills which he brings to his job, but at a deeper sub-conscious level, he is also influenced by elements of his common-sense culture.

Considering basic spatial references in Singapore, primarily public space, this can be defined in the HDB context as the space within the estate but outside the private dwelling unit, that is, the corridor, the lift and stairwell, void deck, carparks, playgrounds and other surrounding areas. The social significance of these areas in public housing estates has been highlighted as well as discussed in some detail and hence, will not be repeated here (Ooi and Tan 1992).

Experience with high-rise, high-density public housing in the Western world has been essentially negative. Many Western urban planners no longer consider high-rise estates as viable solutions to the increasing housing needs of urban populations. The basic criticism of high-rise, high-density public housing focuses almost exclusively on public space.

Pruitt-Igoe, a large-scale "futuristic" public housing project opened in St Louis, Missouri, in 1954, is one such undesirable project. It was reduced to rubble less than 20 years after its construction because it had degenerated into an unsalvageable urban slum. Yancey (1973) who conducted a study in Pruitt-Igoe prior to its demolition, concluded that the inhabitants felt alienated from public spaces which he defined as all space outside the private domain of the individual dwelling unit. Conversely, they extolled the facilities in the individual dwelling units and were also willing to invest time, effort and money in maintaining this private space.

In evolving an approach to research on high-rise, high-density public housing, it must be remembered that the public space is the key to assess the viability of a large-scale public housing project. Thus, identification with public spaces and the mechanics of their maintenance are important aspects of any research agenda on public housing.

Since Western research was written through the perspective of Western norms and values — Western common-sense culture, it seems that the Singapore-housing experience is almost a contradiction as, thus far, it appears to be largely a positive experience. If the Singapore high-rise, high-density experience is not fundamentally different from the Western experience, then the latter will dictate, in the long term at least, an inevitable negative experience. The question to be asked then, is, to what extent does the Singaporean experience differ? This question can be answered partially through the perspective of common-sense culture. Such a perspective describes Singapore public space in terms of what might be culturally meaningful to Singaporeans. Singapore's new, modern, metropolitan form — with its central city core and New Town rings contains some elements of a common-sense culture, which is Eastern rather than, entirely Western.

Chinese cities and villages are highly structured spatial entities with definite boundaries, orientations and functional areas. In order to support such a generalisation, one needs to refer again to the notion of common-sense culture. One can even borrow Levy Bruhl's term, "collective representations" to characterise some fundamental elements of common-sense culture. According to Bruhl, collective representations are powerful but overlooked as cultural norms and patterns of behaviour which are passed directly from generation to generation. Speculating on collective representations prove the very core of common-sense culture. That it is uncharted and controversial does not mean that social scientists should not be encouraged to formulate its possible influences.

Collective representations as complete explanations of common-sense culture cannot, of course, be seen as complete explanations behind the daily activities in modern Singapore. They are only useful to trace behaviour which can neither be attributed to climate nor measured on a low-versus high-density living, urban versus rural life, or modern against traditional lifestyles. For the sake of argument, the relationship between Chinese common-sense culture and the Chinese perception of space might be characterised as territorial compartmentalisation. This principle was also present in pre-HDB Singapore. If one reads, for example, Kaye's (1960) study of Upper Nankin Street in the 1950s, one is struck by the fact that, in Chinatown, there was a precise demarcation of territory into minute cubicles and even bunk space in corridors. This sense of territoriality spilled out onto the streets. Certain streets or parts thereof "belonged" to certain associations, clan groups or secret societies. Such a high sense of territoriality made life at least possible, if not pleasant, under extremely densely-populated conditions.

The HDB concept of walled "courtyards in the sky" or the recent attempts to enclose precinct space with walls and to create walled playground courtyards, evoke elements of the common-sense culture of the Singaporean designer. The spacing of the built form based on this Chinese principle of territorial compartmentalisation and viewed as a collective representation, is characterised by the following. First, as cultivated land in China was scarce and the population dense, the regulation of land ownership became a cultural preoccupation. Second, that intensive land management and use led to an increasing subdivision of individual plots. Third, the Chinese inheritance system supported the perpetuation of land ownership through successive generations. Furthermore, this Chinese common-sense cultural conceptualisation is as relevant for understanding the meaning of Singapore's new towns as it was for the Chinese walled cities of centuries ago.

Singaporean common-sense culture is not only Chinese; it contains contributions from other ethnic cultures as well. As in the case of the Chinese, elements of Malay common-sense culture are operative through several historical periods. This, of course, does not mean that such common-sense culture is static and unchanging. But because it is seldom discerned, just how it changes and under what circumstances, are subjects beyond the present discussion.

Malay culture and its relationship to space contain some elements which differ from that of the Chinese — for instance, the lack of emphasis on territorial compartmentalisation. In Malay *kampung* space, there are no definite boundaries, no sense of territorial orientation and demarcations of functional areas. It could be argued that the relationship between Malay common-sense culture and the Malay perception of space is marked by, what might be termed, territorial obfuscation. There are, at least, three reasons for this. First, Malays were originally island people; thus, trade and subsistence fishing made mobility of the essence. Second, even when Malays became peasants, land had no intrinsic value unless it was productive. New fields could be acquired by simply clearing jungle and houses erected quickly and cheaply. There was, therefore, no need to perceive land as a resource which needed to be "owned" but only which could be "used" by an individual during his lifetime. And third, this lack of interest in rights over territory was reinforced by

Islamic inheritance laws, which discourage the accumulation of land over successive generations.

Take for example, the Perseverance Estate of the 1950s, more commonly known as Geylang Serai. Even in this emphatically urban, highly dense population manifestation, this Malay *kampung* remained true to the principles of non-territoriality. Land ownership was unimportant since houses were built, bought and sold. Pathways meandered through the *kampung* and there were no enclosing fences. Usufruct was respected where the produce of a certain tree itself was emphatically respected.

This issue of Chinese territorial compartmentalisation versus Malay territorial obfuscation poses some vital questions indeed. To what extent can universal human values such as territoriality be considered, and to what extent can it either be conceptualised as profound or superficial? Are there differences in the way in which territories are defined or expressed? Is there a difference in the degree of territoriality? In other words, do the Chinese have a stronger sense of territory as against the Malays? Or does the manner in which they perceive or articulate territoriality simply differ?

To return to a concrete example, the ownership of the Haji's fruit tree is fundamentally different from the concept of land ownership, which is a more permanent entity. The tree, unlike the land on which it stands, obviously has a finite existence. It dies and it can also be cut. Furthermore, it is doubtful if the rights over the tree can be inherited.

The present "no-man's land, everyman's land" status of most public space in the HDB estates could be explained as an essentially Malay common-sense element which has been unconsciously adopted by the HDB administrators. Take the use of street furniture which is liberally strewn in the public spaces of every estate. Who has proprietary use over certain pieces of street furniture? The answer is **no one**. It is not linked, in any way, to a sense of territoriality. The proximity to a certain block of flats, in no way, means that the residents of that block can claim proprietary rights over that street furniture. The Chinatown sense of street territoriality is completely absent in the estates and instead, the Malay concept seems to prevail.

Even the meandering paths which bisect public space are more reminiscent of a Malay *kampung* than that of a well-ordered Chinese urban space.

Singaporeans have built modern Chinese walled cities but they have assigned no territorial proprietary rights to the vast public spaces within. A contradiction here becomes interestingly and bewilderingly apparent, if one is not aware of its commonsensical origins in the clash of Chinese and Malay elements of Singapore "small c" culture. Take the following examples.

Fruit Trees

There are now some 18 000 fruit trees planted within the estates and many of them have begun to bear fruit. Following the Chinese sense of bounded territoriality, fruit trees would probably belong to residents in the closest proximity. Following the Malay principle of usufruct, trees would be assigned owners, not necessarily on the basis of territorial proximity but, perhaps, the customary rights of care givers. This is a very ambivalent situation wherein no one seems quite aware of what belongs to whom. Residents' Committees, through the setting up of gardening clubs, are supposed to tackle the problem. But a *Straits Times* article recently stated that the fewer than 400 people had not kept up with their gardening club credentials and that the HDB should step in. Solving any problem is predicated on understanding its nature. Until one opts for either the Chinese or Malay pattern, fruit trees and their fruits will remain an HDB basket case.

Litter

This concerns residents of Block 28 Marine Parade Estate. Block 28 is situated on the pedestrian thoroughfare which connects the buses along Marine Parade Road and the beaches of the East Coast Park. Residents complain about the litter exacerbated by the many passers-by who discard their rubbish on their way past the block. Such complaints that the litter clutter the public space surrounding their block certainly demonstrate that the residents of Block 28 identify with the unbound public space adjacent to their flats. But because there is neither a Chinese sense of territoriality nor a *kampung* style *gotong royong*, the residents' only solution to their problem is to appeal to the HDB to police the territory, place expensive concrete slabs over the drains to reduce rubbish and to raise maintenance costs by having to engage HDB cleaners more frequently. Again, the ambivalence towards public space-use and the absence of clear guidelines on territoriality have led to a vacuum that must be administratively and institutionally filled.

In conclusion, it might be relevant to speculate upon certain similarities. Although the attitude towards territory seems to differ, both Chinese and Malay attitudes share a common insistence on the importance of a socially constructed principle of bounded space *per se*. Thus, the Malay principle of *gotong royong*, whereby public space within a particular *kampung* was maintained by the residents, was also predicated on a certain control over access to and use of this public space. A stranger entering a Malay *kampung* would have been quickly identified. The open verandahs of traditional Malay houses served as "police posts" where the presence of strangers was quickly registered. This system is also evident among the Chinese. One reads, for example, of Chinatown gang violence, which takes place when the territory of one gang is trespassed, sometimes, by a innocent stranger.

The HDB public space is unbounded in the sense that access to any estate is open. Efforts to contain littering can only be fruitful when access to the space is controlled so that individuals who enter can be held accountable. Residents, as in a *kampung* would, perhaps, be more interested in maintaining their public space, if they also had some control over access to this space.

Time

Time and space are usually viewed as parallel dimensions into which activities and objects must be slotted. While these may be intrinsic to the notion of space, time is another matter.

Time does not necessarily have to be viewed as a continuum upon which objects and activities are plotted in linear fashion. The Western concept of time or what might be termed "clock time" imposes order upon objects. This orientation needs to be looked into in the context of the common-sense culture of Singapore. Supposing this Western concept of time is inverted and it is assumed instead that objects and activities impose themselves on time, not vice versa. So, suppose that time does not regiment activity, but rather, activity regulates mealtimes and fasting times affect daily routines. The Muslim fasting month, for example, literally inverts day and night schedule whereby no food or drink is consumed between sunrise and sundown and therefore, the night hours are used to cook and eat. Similarly, certain rituals tied to the Chinese calendar have been imposed on the Western progression of time. There are concrete examples to illustrate this point.

Wet Market Time

The Western development of clock time which divides the day into 24 equal hours is a rather recent phenomenon. One could hypothesise that it originated with the advent of wage labour, where a person is paid for the number of hours he worked. The longer the number of hours worked, the higher the productivity and the higher the wage. Thus, the idea of productivity became linked to clock time. One of the major criticisms of the Singaporean phenomenon of wet markets, many of which are in the HDB estates, is that they not only waste space but also time. Wet markets only operate about five hours a day, from 7 am to 12 pm. Such short operating hours are wasteful, it is argued, in that five hours of productivity is an insufficient time investment in labour. If wet markets are converted to small sundry shops, lock-up stalls or mini-markets, the argument goes, they can remain open 12 to 14 hours a day.

The argument, of course, does not take into account the obvious fact that stall owners have actually begun work since 4 am at least, collecting and arranging the fresh produce, and work extra hours after closing time, packing and storing away their remaining produce for the next day. Leaving this argument aside for the moment, consider what happens if the merits of wet markets are measured not in Western clock time but in Singapore's common-sense culture time.

First, in Singapore's tropical climate, marketing activity dictates morning time not only because it is more pleasant to market in the cool of the morning, but because, otherwise, market time must be air-conditioned time, given the perishable state of fresh food products. Marketing activity, therefore, dictates a short-time scale, if it is not to incur higher costs.

Second, and perhaps more important, are the shorter operating hours. The management of the smaller, more specialised market stalls selling only several kinds of vegetables and other market produce, which requires only several operating hours a day, is ideally suited to those who need a small income but do not wish or find it impossible to work the longer periods of 12 or 14 hours a day that the lock-up stall or mini-market would necessitate.

Continuing the relatively light labour intensive activities which have been a traditional part of Singapore labour force activities, seems important because of the number of active and young retirees who will be an increasing proportion of the HDB population. If all productivity were dictated by Western time life cycles, then many traditional sources of income generation would probably disappear.

Eating Out

The second example concerns eating. Here again, in Western cultures, eating is dictated by clock time. In Western Europe, if one does not eat according to clock time, he will not eat at all. It is almost impossible to have anywhere in Spain, a late lunch at, say, 3 pm, because that is already *siesta* time.

What about Singapore's common-sense culture time? With regard to eating, there are two states of being which alternate throughout the day and also well into the night. This is, before eating and after eating. These two states have been routinised into a greeting — one does not say "hello" but "have you eaten?". In Singapore, the activity of eating in response to hunger determines eating time. There is never a time when someone is not eating noodles.

Any change in our eating patterns and any shift to something more regimented would certainly have grave implications on Singapore's hawker industry. Some hawker centres are favoured as gathering places in the mornings and early afternoons. Others cater to late-afternoon and early-evening crowds while there are centres that remain open for those who need a midnight snack. If the European mealtime schedules were imposed here, then hawker centres would have to be re-designed to cater to floods of people at regimented times of the day and nobody for a stretch of hours. This would, in turn, certainly affect the quality, character and variety of the food prepared.

In the Singaporean context, environmental change obviously influences cultural change, but common-sense culture also influences environmental change. Isolating and articulating elements of common-sense culture make for an important task. Understanding its applied research potential is crucial to the maintenance and continued improvement of the quality of life.

REFERENCES

Chermayeff, S. and C. Alexander (1976) *Community and Privacy — Towards a New Architecture of Humanism* (Garden City, New York: Anchor Books).

Kaye, B. (1960) *Upper Nankin Street* (Singapore: University of Malaya Press).

Lynch, K. (1980) *A Theory of Good City Form* (Cambridge, MA: Massachusetts Institute of Technology).

Ooi, G.L. and Tan, T. T.W. (1992) "The Social Significance of Public Places in Public Housing Estates", in B.H. Chua and N. Edwards (eds.) *Public Spaces — Design, Use and Management* (Singapore: Times Academic Press), pp. 69–81.

Relph, E. (1976) *Place and Placelessness* (London: Pion).

Tan, T.K.J., Loh, C.T., Tan, S.A., Lau, W.C. and Kwok, K. (1985) "Physical Planning and Design", in A.K. Wong and S.H.K. Yeh (eds.) *Housing a Nation — 25 Years of Public Housing in Singapore*, Maruzen Asia for Housing Development Board, (Singapore).

Yancey, W.L. (1973) "Architecture, Interaction and Social Control — The Case of a Large-Scale Public Housing Project", in J. Helmer and N.A. Eddington (eds.) *Urbanism — The Psychology of Urban Survival* (New York: The Free Press).

8

Lack-of-Fit: Effects of the Public Housing Programme on the Poor and the Elderly

Myrna Blake

Introduction

Affordable housing is essential to the individuals and families who have to cope with low incomes. In Singapore, the provision and management of public housing is the mandate of the Housing and Development Board (HDB), a statutory body established in 1960. In these thirty odd years since its establishment, the HDB has provided affordable housing to 86 per cent of Singapore's population of some three million people. The achievement is nothing short of phenomenal.

Two features of the public housing programme have special significance to the poor. The first relates to the assumptions which the HDB adopts in order to achieve its impressive results. The second is the administrative structure necessary to keep pace with the development of the public housing programme.

The assumption of a family profile is necessary in the process of housing over 620 000 family units. This profile involves a two-generation nuclear family consisting of husband, wife and two children and includes an older generation of parents living together with the two-generation family or in close proximity, with the various units being in frequent contact, at least by telephone.

All the members in this "normal" family are presumed to be physically and mentally fit, upwardly mobile and aspiring not only for an HDB flat, but a larger one. In this family, the husband is the main wage-earner; he is regularly employed and accumulates substantial savings in his Central Provident Fund (CPF). He holds a bank account, probably with the Post Office Saving Bank (POSBank) and pays his bills through an automatic transfer (GIRO) system. The wife may also be employed outside the home but is considered a secondary wage-earner, supplementing the family income. Unless child-care costs are affordable, she may give up her job in order to look after the children herself. The wife, if not some other individual, is assumed to be available to care for her children and the other dependents such as a disabled elderly parent, or other family members with some form of impairment.

At least one member in this family is presumed to be literate. This person is likely to be a parent between the ages of 30 and 45, or an adolescent.

Indicators from various studies are fed into the development of this profile so that it does, in fact, apply to a majority of Singapore families, especially those living in public housing. 80 per cent of these families own the flats which they live in. Many of the poor, however, do not fit this norm, and this "lack-of-fit" engenders problems for housing.

The second significant feature of public housing for the poor relates to the first. A bureaucracy characterises the public housing programme. The planning, building and administration of more than half a million residential and other built units require a sizeable institution, which the HDB has become. The then Chief Executive Officer of the HDB, Mr Liu Thai Ker, admitted in 1990 that, several problems have stemmed from the "massive size of the board" and "red tape in the form of ever-growing rules and regulations is strangling work efficiency and slowing the service to the public" (*Straits Times*, 14 March 1990).

It is difficult to locate precisely, the cut-off point of poverty in Singapore. The National Council of Social Services (NCSS) has established an administrative guideline which makes it possible to define poverty in absolute terms. For the average Singapore family of four persons, a monthly total income of S$640 and below would qualify a family for the Rent and Utilities Assistance Scheme, thereby defining the family as poor. Based on the 1990 Census data, it is estimated that approximately 3.6 per cent of all households of four or more persons live on incomes which fall at or below the NCSS poverty line. However, even a monthly total income of slightly above this line would make a family vulnerable to poverty, since there is no margin of savings. This vulnerable group called the "underclass" has been estimated to constitute about 10 to 15 per cent of the population (*Straits Times*, 26 February 1990).

Given the home-ownership aspirations embodied in the norm, it is reasonable to expect residents in one- and two-room flats to include a large proportion of the economically vulnerable. It is a mistake, however, to assume that the underprivileged are exclusively contained in this group or that all residents of the one- and two-room flats necessarily manifest poverty.

Experience of voluntary welfare organisations concurs with the findings of the Advisory Council on the Family and Community, namely, that families living in poverty fall into one of the following categories — single parents with dependent children and especially, when the parent is a mother; the principal wage-earner suffering from chronic disease or physical, intellectual or mental disability; elderly individuals or couples living on their own without means of support; the principal wage-earner is deceased; parents, particularly, the principal wage-earner, suffer from personality disorders which might lead to imprisonment or detention in drug rehabilitation centres.

Since the principal wage-earner in a family is, generally, a male, most of the above factors place heavy burdens on women. The earning capacity of women with little education or skills is usually low. When the necessity to look after young children limits the time during which she can work outside the home, her earning capacity is reduced even further. Her take-home pay rarely exceeds S$500 per month. Families supported by women in these circumstances experience extreme difficulty in meeting regular payments for rent and utilities, in addition to the families' daily expenses.

The major emphasis of this paper is on the outcome of economic difficulties of the poor in relation to the administrative aspects of public housing. The discussion on the frail elderly will focus on some physical features of public housing and other economic details.

The Built Environment of Public Housing

When the HDB was established in February 1960, it adopted as its immediate short-term plan, "a programme to build as many low-cost housing units as possible to meet the needs of people in the lower-income groups whose requirements have never been catered for by private enterprise". Linked to the plan was the objective of relieving the acute congestion in Central Areas through the creation of housing estates which would be within easy reach of the city centre (Housing and Development Board (1960) *Annual Report*).

In implementing the plan, the target for the following five years was to build 10 000 units per year. Much attention was paid to the economic capacity of low-income groups and part of its resources was invested in planning and building one-room flats which could be rented at S$20 per month.

In the last 30 years, the HDB has built some 600 000 residential units. Growing affluence in Singapore has stimulated the HDB to extend its original objectives to the provision of higher standards of housing for middle-income groups and to pay attention to the aesthetics of buildings and estates. However, the need for low-income housing, principally the one- and two-room flats has persisted.

Table 8.1 summarises the developments in the last three decades. The figures in 1961 reflect the activity and policy of the Singapore Improvement Trust, precursor to the HDB. Since 1971, the one- and two-room flats under the HDB management have been reduced from 58 per cent to 11 per cent of the residential units. The statistics indicate not only the growing capacity of Singaporeans to upgrade their standard of living but also the diminishing proportion of the underclass or underprivileged — the group most in need of low-cost housing.

Among the households living in one- and two-room flats, a high proportion (96 per cent and 86 per cent respectively) rent

Table 8.1 Distribution of Residential Units Under HDB Management in the Last Three Decades

Type of Flat	1961 (%)	1971 (%)	1981 (%)	1991 (%)
1-room	9.1	34.0	19.2	4.6
2-room	36.9	24.2	13.5	6.7
3-room	46.0	39.2	46.6	40.2
Others	8.0	2.6	20.7	48.5
No. of Units	26200	126700	348900	627100

Source: Housing and Development Board *Annual Reports*, 1960–1991.

them. Together, these households in small apartments comprise ten per cent of the population living in public housing. Not all in this figure can be assumed to be poor. However, considerable importance has been placed on home ownership among Singaporeans and there is apparent motivation among the people to upgrade their standard of housing when they can afford to do so. Therefore, it is likely that the category of the underclass, which is the economically vulnerable group, includes the majority of the 10 to 15 per cent of the population still living in the smaller public housing apartments.

It is an indisputable fact that the public housing programme has improved the living conditions of the poor in Singapore. Conditions such as those described in 1960 by Kaye (1959) were intolerable. The first fully elected government of the time rightly established as one of its major goals, the re-housing of the masses who lived in sub-standard urban tenements and sub-urban and rural squatter areas. Aesthetic limitations provided better sanitation, ventilation, direct and indoor access to a water supply and in some, a healthier living environment. The fact that the major health problem among school children has in the last 30 years, shifted from worm infestation and gastro-intestinal diseases to obesity attests to a growing economy as well as to a better environment.

In spite of the social and economic gains which have accrued to the majority, however, a small minority is not yet able to benefit fully from the opportunities available. This is largely because of a combination of deficits, the most important of which is economic.

Some Aspects of Lack-of-Fit

The following discussion illustrates the relationship between the administrative aspects of the public housing programme and the poor. Case examples demonstrate the effect of the lack-of-fit between underprivileged families and three aspects of the fundamental assumptions on which policies are fashioned. These are the assumptions of regular employment; a certain level of literacy and a complete family in which the division of labour between the male head of household and his wife is complementary.

One aspect of the norm assumed is partly prescribed by the terms governing the purchase or rental of flats. The contract involved in these transactions requires that the total family income does not exceed a stated amount per month at the time of application (S$800 for rental and S$8000 for purchase — a sum raised from S$7000 in 1994 and from S$5000 in 1992). The subsequent contract then necessitates regular payments of rent or instalment repayment of a loan procured for the purchase of an apartment plus the conservancy charges. For everyone's convenience, these payments are made preferably by the system of GIRO, through a bank account and, most commonly, the POSBank.

Debts

One manifestation of the lack-of-fit between the poor families and the assumed norm is the violation of payment terms. At any one time, a number of tenants and lessees are in arrears of rent, conservancy charges, or repayment of loans towards the purchase of an apartment. In several of these families, income is either very low, or is derived from casual or irregular employment. Both of these circumstances are typically linked with low levels of skills and education.

Whether or not the household is headed by a male, special strategies are required for subsistence on low income. Families experiencing economic hardship, sometimes, manage the household budget by alternating the types of bills paid, that is, in one month, they pay the rent and in the next month, the bill for utilities. Inevitably, they fall into arrears and might borrow money to make the necessary payments. This solution merely shifts the debts from one location to another. These debts increase the demands on the household budget. In this situation, immediate needs like family expenditures on daily expenses, medical bills, school expenses, aggressive demands for repayment of debts, and the threat of disconnection of water or electricity supplies take priority.

To overcome the problem of arrears to the organisation, the HDB officers work out an instalment plan with the family. Sustaining these payments on a low income is a struggle. Without major change within the family structure like the addition of another wage-earner, support provided by members of the extended family or assistance from external sources, it is sometimes impossible for the family to escape the debt cycle.

Irregular Earnings

Low income is often linked to irregularity of earnings. While an irregular income might be the result of the availability of work for individuals with low levels of education and skills, it is sometimes the outcome of rational decision-making regarding the best means to earn an income. Regular jobs for those with limited education and skills, pay very little and compulsory contributions to the CPF reduce the take-home pay. It therefore makes more economic sense to the wage-earner to take on contract work or part-time work on a daily basis. An example of this decision is illustrated in the following case:

> Case 1
>
> Mr. Quek, 28 is the principal wage-earner in a family consisting of his parents and two younger siblings in school. His father is terminally ill and unfit for work. His 50-year-old mother is employed as a cleaner, earning S$13 a day. Mr. Quek could have obtained regular work as a factory worker. However, he calculated that, after deduction of CPF contributions, his take-home pay would have been approximately S$400 per month. Furthermore, the work means

> daily drudgery. Therefore, he opted to work as a daily-rated contract labourer. He earns S$30 per day and more with overtime work. After two weeks of such work, he takes home S$600. He may then enjoy a period without work. Unfortunately, the irregularity of this income often means that, by the time he earns S$600, he and his family have accumulated several debts to which most of the income will go to as payment.

Another advantage of irregular work in the form of daily-rated employment is the frequency of income payments. When a family lives from hand-to-mouth, there is often no margin of savings. After the monthly bills are paid, a small salary evaporates. In these circumstances, money in hand, even if it is only S$25 per day, allows a cash flow which a monthly wage payment does not.

A third advantage of irregular work is the time available between jobs. The struggle to live on a low income requires a great deal of crisis-averting activity. When accumulated debts create a potentially explosive situation like when the family is threatened with eviction, utilities are about to be cut off or loan sharks become aggressive, then immediate action has to be taken to temporarily stave off the consequences. This might mean finding other sources of money or negotiating with authorities. All these steps take time. If the resident is employed in a regular job, a day off just to take care of outstanding debts may result in a black mark, a reprimand or even a deduction from wages. Daily-rated work on a contract basis allows for greater flexibility.

The HDB has not been unresponsive to the needs of the poor. The concern for low-income groups is evident in the availability of rental flats at low rates, S$40 for one-room and S$70 for two-room flat. However, the lifestyles of those living on low income are, perhaps, insufficiently appreciated. The requirements of regular payments and the imposition of penalties for default add to the difficulties of families struggling to survive on low and irregular incomes.

Low Levels of Literacy

Another assumption underlying the operations of the HDB is a level of literacy which permits comprehension of formal letters and notices. In the relationship between the HDB and its tenants and lessees, communication is essential, especially when routines do not go according to plan as, for instance, when required payments are delayed. In such instances, three communication procedures are as follows:

Beginning from the second month for which rent is not paid, a tenant will receive the following letters from the HDB — a reminder letter, a notice to quit which warns about the prospect of legal action and eviction, a letter of demand which precedes the implementation of legal action, a writ of summons indicating the initiation of legal action, a notification of judgement which informs the tenants of the outcome of court proceedings, and a writ of possession which empowers the HDB to repossess the flat in question. This communication also sets a date for eviction of tenant.

For persons who default on repayment of loan instalments or of conservancy charges, the procedures are almost similar. Several written reminders are sent prior to a notice demanding immediate payment. Then a final notice is served and ultimately, a notice of compulsory acquisition.

In conjunction with the written communications, the HDB officers also attempt to visit the defaulters and to explain procedures and consequences. The whole process may take several months partly, because of the heavy workload of the HDB officers as well as the attempt to give the defaulting tenant a chance to settle accounts.

From an administrative point of view, the system is fair and allows a great deal of time and opportunity for settlement of debts. However, for the tenant with little or no education, the process is confusing. Although the letters are written in four languages, the terminology and phrasing in the letters are legalistic and formal. Recipients with little education do not fully understand the letters. Furthermore, reminders and other important notices are often ignored because many people would have heard from their friends and neighbours that nothing immediate happens consequent to the letters.

Difficulties also accompany verbal communication. The visit from the HDB officer might not fully convey the seriousness of the consequence of sustained debts. When a person with little education attempts to negotiate with desk-officers at the HDB branch office, he often lacks the necessary skills to ask appropriate questions and to explain his or her circumstances or to grasp what is required. The combination of over-burdened front-line officers and supplicants under stress frequently results in a flaring of tempers, threats and hardening of positions. Chances of rational negotiation are jeopardised. Defaulters are reluctant to return to the area office to re-negotiate and their problem is compounded. When no solutions are negotiated, the consequences for the family can be painful. The following case illustrates one such example:

> Case 2
>
> Mr. Joseph bought his three-room flat after living in a one-room flat for a couple of years. Despite an erratic employment record, he accumulated enough money in his CPF to make him feel that he could afford to purchase a resale flat. His income of S$550 was barely sufficient to support his family of five and he was unable to sustain the instalment payments on his flat. He did not understand the various notices from the HDB and frequently ignored them. When he visited the HDB branch office, his explanation and appeals were refused and immediate full payment was demanded. His own aggressive manner probably contributed to the demanding responses from the desk-officers.
>
> When convinced of the seriousness of the situation, he did attempt to make payment. By then, the legal process were irreversible. His flat was reclaimed by the HDB and he lost the CPF savings that had been paid towards its purchase. Though he received monetary compensation for the re-purchase of the flat, he continued to owe the Board part of the loan. He would have been forced to move into a one-room

flat with his family if some help had not been found for him through a social-service agency.

Households Headed by Women

Another factor contributing to poverty is that of single parenthood — especially where the single parent is a mother. However, one assumption embodied in the HDB administrative procedures which affect the poor is that of a complete nuclear family with a male household head. When the man is the registered tenant of an HDB flat, his wife cannot terminate an agreement between her family and the HDB unless she can produce documentary evidence of widowhood, a divorce or a legal separation.

Similarly, if she is a co-lessee of a purchased flat, any change in an agreement with the HDB must be made jointly between the husband and wife — a practice that can add to her difficulties. Two examples of these difficulties that have been experienced by women are illustrated as follows:

Case 3

Madam Lee, 32, had a husband who set up a second household with another woman. Her husband returned to his marital home occasionally and erratically. Unable to cope financially, she decided to convert from a two-room to a one-room flat. However, the tenancy of the two-room flat was in her husband's name and she could not make the change without his written consent and he refused to terminate the tenancy. While he did make an attempt to settle the arrears, his maintenance continued to be irregular, despite a court order. Madam Lee and her three school-going children were left trapped in an untenable situation. The debarment could have been removed had there been proof of divorce and/or all debts cleared.

Case 4

Madam Chia, 42, was deserted by her husband ten years ago when her children were 12 and 14 years old. Several years later, when the HDB was relocating residents from the area in which she lived, she was told that she had to move. However, because the flat had been registered in her husband's name, she was unable to have it transferred. By this time, she had lost contact with her husband and had no desire to keep in touch with him.

Other Effects of the Public Housing Programme

Administrative procedures exacerbate but are seldom the sole cause of problems besetting the poor. Family and personality dysfunctions are not uncommon, though, not exclusive features among the underclass. These generate difficulties in a variety of contexts as several studies have shown (Lewis 1959; Minuchin 1970; Madge 1983; Wilson 1987; Macleod 1987). However, certain characteristics of the Singapore environment make for particular problems.

The first is a lack of alternatives. The success of the HDB in clearing sub-standard housing areas has meant the almost complete annihilation of the areas and individual dwellings where more informal relations between landlord and tenant were possible. While disadvantageous to the tenants in some respects, these arrangements allow poor families to use a varicty or strategies for survival like paying enough of a debt whenever possible to maintain evidence of seriousness to settle accounts and thereby, to delay retributive action. Then, there are ways of obtaining electrical and water connections while avoiding payment of public utilities.

A second aspect of the public housing policy disadvantageous to the low-income groups is that the amount of space allotted to a family is related to the family's ability to pay rather than its need. The situation of a family of seven in a one-room flat of approximately 350 square feet is not conducive to both good-child development and family harmony. Though a narrow access corridor outside the flat constitutes extra space in which children may play, the high-density living conditions within the flat are not satisfactory for school children who wish to study.

Furthermore, adolescent peer-group activities in void decks and other common spaces in the estate, frequently attract teenage children who find the closeness of family restrictive. Such a condition combined with the low income of these families results in a situation wherein access to the only available or affordable form of housing means that the options open to the poor have, in fact, been reduced.

Policy Implications

In terms of policy, the above discussion points to a need for greater flexibility in the operational rules of the housing programme and a great deal of assistance to selected families so that they meet the terms of eligibility for public housing. In fairness to the HDB, it must be noted that a certain amount of flexibility is embodied in the day-to-day operations with some delegation of decision-making powers given to the HDB branch officers. Attention to refining the means of assessing economic vulnerability would enhance their capacity for flexibility.

Adequate risk analysis and screening of applicants for the purchase of flats such as those conducted by a commercial bank when clients apply for loans, will reduce the chances of future debts. More time and effort could be allocated to explaining the purpose and consequences of various administrative processes so as to ensure full compensation from tenants and lessees. This may well be effective in moderating, if not, preventing the accumulation of debts, as well as the conflicts and misunderstanding that arise between the HDB and its clients.

Housing Design Features: Effects on the Elderly

There has been an understanding that the first targets of a public housing programme are the needs of the masses or the majority. If such a principle becomes the basis of planning, the few who do not fit into the profile of the majority are accordingly disadvantaged. One such group is the elderly. This group is handicapped by several physical aspects of public housing design.

The first of these is the HDB building practice of providing a lift stop on every fourth floor of a multi-storey building. This is disadvantageous to any person, young or old, with a physical handicap. An elderly person who is hemiplegic as a result of a stroke, might achieve some proportion of mobility with a walker. However, mobility outside the home is limited if he or she does not live on a floor with a lift stop or if he or she lives in a four-storey walk-up block. The need for help severely decreases the older person's independence, an attribute which is important to a person who, in the course of ageing, would have experienced many other losses. The need for help shrinks the options available to the elderly person and can result in the deprivation of services which could have increased their independence.

Recent public discussion on the needs of the aged have resulted in an increased awareness of the problems. The HDB plans for future blocks of flats to include lift stops on every floor and to do away with four-storey walk-up blocks. The accessibility of the buildings for physically handicapped persons in wheelchairs is addressed. Special parking lots have been allocated and ramps have been installed in a few places.

Uniformity of the blocks in a housing estate is a second feature of public housing which creates difficulties for older people. An elderly person who moves into a new estate with a young family is often disoriented. Such a state of mind arouses an anxiety about going out and not being able to find the way home. When the young members of the family go out to work, the old persons are left alone at home and this fear of getting lost isolates them for many hours.

When mental disability attends ageing, a shift from one environment to another, as happens during relocation, can have other effects. Old people who have moved from more informal environments of the *kampung*, sometimes, find the new housekeeping requirements rather daunting. The drainage, plumbing and garbage systems require a different system of maintenance. When old people find it difficult to change methods of housekeeping, the results can be disastrous even to their neighbours. Fortunately, however, these difficulties will diminish in the future, as there are only a few thousand households which have yet to face the process of resettlement. In the meantime, some elderly persons still experience problems associated with resettlement as in the following case:

> Case 5
>
> Che Rahim, 89, lives with his 85-year-old wife who is mentally ill. They share their one-room flat with several cats. The food left for these cats attracts cockroaches. The couple is unable to fathom the complexities of the plumbing system and their drains are frequently clogged, resulting sometimes in the flooding of their floor. The stench of rotting garbage and invasion of cockroaches create not only unwholesome living conditions for them but also an irritant for their neighbours, who are, therefore, not inclined to friendliness.

The development of new, modern and improved housing estates in the suburbs of Singapore has other implications for the older generations of residents and particularly, among the low-income groups. The younger generations and more affluent generations have moved to the new areas to avail themselves of the improved facilities. This movement has left the elderly, particularly, those with low incomes, in relatively high proportions in the older estates, especially where one- and two-room flats are still available. These concentrations which are visible in certain housing estates are almost similar to ghettos.

However, it is also found that the natural consequences of long-term residence in one area produce a phenomenon which ameliorates the impact of the emigration of young people from the community. For instance, it is found that social links and support networks develop among the elderly in the Malay community. The individuals meet regularly, introduce each other to various social services, help each other as in collecting food rations, and they feel the common bond of being on "welfare", hence, reducing the stigma of the condition (Blake 1992). It is highly possible that similar cases develop among other ethnic groups. There are instances of friendly and generous hawkers who, because of familiarity through the years of living with long-term residents, recognise the elderly poor who live on their own and know of their difficulty managing on low incomes. Frequently, these hawkers have given free or discounted meals to the elders that they know of.

Conclusion

The public housing programme in Singapore has helped improve the living standard of families who live on low incomes. However, the bureaucracy with which the programme is encumbered, sometimes dilutes the benefits. The terms associated with housing are based on assumptions which do not apply to a small minority of the population. For these people who cannot easily fulfill the conditions necessary for access to public housing, there are few alternatives. While social effects of the built environment are not always anticipated, spontaneous social processes sometimes moderate the consequences of a building programme. Still, this does not reduce the responsibility of builders and administrators to sharpen their capacity to respond to the needs of the most vulnerable and least vocal sections of the population.

NOTE

1 In a recent study that I conducted, it was found that 270 out of 636 lower-income families lived in one-room flats. Of these, 20 households consisted of 7 or more persons.

REFERENCES

Blake, M. (1992) *Growing Old in the Malay Community*, Center for Advanced Studies, National University of Singapore, Occasional Paper No. 10, (Singapore: Times Academic Press).

Department of Statistics, Singapore (1991) *Census of Population 1990, Households and Housing.*

Housing and Development Board *Annual Reports*, for the years 1960–1994.

Kaye, B. (1959) *Upper Nankin Street* (Singapore: University of Malaya Press).

Kelly, R.F. (1985) "The Family and the Urban Underclass — an Integrative Framework", *Journal of Family Issues* Vol. 6 (2): 159–184.

Kelly, R.F. (1989) "The Urban Underclass and the Future of Work-Family Relations Research", in E.B. Goldsmith (ed.) *Work and Family — Theory, Research and Applications* (California: Sage Publications).

Kuo, E. (1975) *Families Under Economic Stress* (Singapore: Institute of Southeast Asian Studies).

Lewis, O. (1959) *Five Families — Mexican Case Studies in the Culture of Poverty* (New York: Basic Books).

Lewis, O. (1968) *A Study of Slum Culture — Backgrounds for La Vida* (New York: Random House).

Macleod, J. (1987) *Ain't No Making It — Levelled Aspirations in a Low Income Neighborhood* (London: Tavistock).

Madge, N. (ed.) (1983) *Families at Risk — Studies in Deprivation and Disadvantage* (London: Heinemann).

Ministry of Community Development (1989) *Report of the Advisory Council on Family and Community Life*, May, Singapore.

Minuchin, S. (1970) *Families of the Slums* (New York: Basic Books).

Salaff, J. (1988) *State and Family in Singapore — Restructuring a Developing Society* (Ithaca: Cornell University Press).

Wilson, W.J. (1987) *The Truly Disadvantaged — The Inner City, The Underclass, and Public Policy* (Chicago: University of Chicago Press).

9

Ethnic Relations and the Management of Ethnic Differences in a Residential Community

Lai Ah Eng

Introduction

The management of ethnicity and ethnic relations in the residential setting is a largely unstudied aspect of Singapore's highly acclaimed public housing programme. This is surprising, considering Singapore's multi-ethnic society and the explicit and deliberate aim of the public housing programme to integrate the different ethnic groups through shared residential living, which commensurates with the political ideal of promoting a national identity based on "multi-racialism". Indeed, the large-scale multi-ethnic residential communities created by this deliberate design is historically unprecedented in Singapore, affecting nearly 90 per cent of its entire population which had tended to cluster together in ethnically more homogeneous settlements in the past.

In the heterogeneous and highly-planned context of the public housing estate, the formation of a multi-ethnic community and specifically, the management of ethnic differences is important for two parties — the residents and policy-makers. For residents who have no other alternative but public housing, living in such a multi-ethnic environment simply has to "work". Living together on a daily basis as neighbours and co-residents involves the evolution of a community that is at least tolerable but which, at its best, can be thriving with activities and exchanges, in spite of and because of ethnic differences.

For policy-makers, the political credibility and claim to pragmatism hinge on the success of their ethnic integration strategy as well as on public housing estate management. This link is particularly significant in the light of recent changes in both areas of policy and estate management. In the current context of growing open discourse on ethnicity and ethnic relations in Singapore, understanding this arena of social life may provide useful insights for both residents and policy-makers alike.

This paper examines how some ethnic differences are managed by both parties — the residents[1] and the Housing and Development Board (HDB) at the local residential community level. Specifically, it focuses on cultural-religious differences via selected ethnic occasions which occur within the locality of Marine Parade[2], a residential estate where the policy of ethnic integration has been implemented. Through such a micro-level study, the subtle and complex processes and negotiations operative in the continuous construction of the multi-ethnic community may be better understood and appreciated.

The focus on ethnic differences does not reflect any preference for or emphasis on these over similarities or other emerging forms of ethnic relations. On the contrary, as will be discussed, emphasising similarity is one way of coping among many, in facing and resolving ethnic differences.

The significance of "difference" lies in its potential for dispute and conflict which require management and resolution for community life to be viable. Here, it should be emphasised that reference to "Chinese", "Malays", "Indians", "Eurasians" and other ethnic categories does not necessarily assume homogeneity within each. This study recognises that each ethnic community is highly heterogeneous in their linguistic, cultural, religious and class domains and in fact, the special occasions themselves often reveal this heterogeneity. However, these internal differentiations are beyond the scope of discussion here.

Special occasions are important in the ethnically heterogeneous context because they provide the opportunity for individuals and groups which are normally segregated to interact (Frankenberg 1978). They are "special" not only in that they are, like rituals, highly patterned symbolic meanings, and which break and set them apart from the ordinary flow of everyday life. Their "specialness" also arises from their "alerting quality" through rules which bound and mark them out as significant and therefore, require alertness, attention and appropriate observance (Lewis 1980). Their regular occurrences within the local community's regular and annual lifecycles render them part of its "public" life in two senses. First, they are public in that they are recognised and transmitted socially, independent of the individual who performs it. Second, they are public in their spatial execution. They are, therefore, constituent parts of the community's social and cultural environment which, given the shared setting, its residents must involve themselves in or respond to, in one way or another, whether as actors or spectators, performers or audiences and "givers" or "receivers".

Yet, their rituals and meanings are not necessarily shared by all in the multi-ethnic setting.[3] Thus, if they say something about the essence of one ethnic community, what do they say to non-members, and what do these others, in turn, think of them and the community represented? Do people always understand the cultures of others in their own terms and interpretative constructs or do they attempt to cross boundaries? Do they judge on the basis of appearance or do they comprehend meanings behind the symbolic forms? What notions and stereotypes of each community emerge from inter-ethnic contact at these rituals? What similarities and contrasts are confronted and comparisons made, and how are contrasts and conflicts of meaning dealt with?

Cohen (1985) provides a useful approach to such an understanding of special occasions through the concept of "symbolic boundary". According to him, rituals possess two qualities — they allow the experience of a community and at the same time, act as symbolic markers of boundaries in relation to others. The efficacy of ritual in boundary maintenance lies in its creation of "a sense of belonging, of identity — and by the same token, of difference from others" (Cohen 1985, p. 53). Cohen emphasises the oppositional character of the symbolic construction of community at its boundaries — "the boundaries are **relational** rather than absolute, that is, they mark the community **in relation** to other communities" (Cohen 1985, p. 58). It is this relational character of ethnic rituals that is of concern here. In short, through the experiences of the diverse events in a multi-ethnic setting, cross-cultural exchange through this relational character of rituals, specifically the management of differences, may be better understood.

The main ethnic special occasions which occur in Marine Parade estate are the Malay events of weddings and Hari Raya Puasa to end the month of fasting, the Chinese Seventh Month Ghost Festival as well as the funerals and occasions involving ancestors, deities or spirits which are usually observed through "pavement prayers". Because of their public execution and expression, they are unavoidable and are, therefore, familiar to virtually every resident.

> I have always lived here, grew up here so I am quite familiar with Chinese funeral, Malay wedding, the auction, the ghost festival. I have ever attended a Malay wedding. (*Neeraj*)
>
> Before, we were familiar with Chinese festivals, we always see the things they put when we pass by the Chinese villages. The only difference now is that they take place nearer to us, sometimes just next door! (*Badariah*).

Such familiarity and cultural socialisation are developed through participation as curious spectators, friendly neighbours and invited guests, with the most intimate forms expressed through invitations, visits and gift-giving. The focus here is on the public relations between actors, performers or hosts and guests, spectators or co-residents on these occasions.

The Multi-ethnic Community of Marine Parade Estate

Marine Parade estate is a typical residential community in Singapore — planned, constructed and managed through the government's public housing programme under the HDB. Its 8000 odd households live in rental (two- or three-room) or purchase (two-, three-, four- and five-room) flats in high-rise blocks of various designs and also varying heights. Facilities at the neighbourhood and town centres provide a considerable level of self-containment with shops, stalls, market and hawker centres to ensure the residents of the basic array of goods and services necessary for daily life. These are further supported by various other non-HDB facilities such as shopping complexes, shops and stalls in adjacent areas. In addition, institutions such as kindergartens, primary and secondary schools within and around it provide for its children's education. The vast majority of residents, however, find their sources of livelihood outside it.

Built on land reclaimed from the sea, Marine Parade's first residents, many of them on resettlement programmes, moved into the estate in stages between 1973 and 1976 through a system of balloting (initially) and selection (subsequently) of flats. It was a logical and popular choice because of its proximity to the Geylang Serai, Katong, Old Airport Road, Joo Chiat, Bedok and Eunos areas where most of them previously lived. Most of these areas may be regarded as ethnic settlements or had large concentrations of certain ethnic populations such as Kampung Eunos (Malay), Katong and Joo Chiat (Eurasians and Chinese Babas of mixed Chinese and Malay origins) and Old Airport Road (Chinese). The result was that Marine Parade's original population was multi-ethnic in composition — a feature that has continued notwithstanding recent movements of residents in and out of the estate.[4] At the end of 1988, its population comprised 70 per cent Chinese, 20 per cent Malays, and 10 per cent Eurasians, Indians and other ethnic minorities.[5]

The making of the multi-ethnic community is premised upon the sharing of a common residential setting which covers virtually every aspect of public space. Living next door to one another or on the same floor or block, neighbours share common corridors, stairways, lifts, void decks[6] and passageways. Residents also share common parking lots, paths and streets, corners and playgrounds, shops, hawker centre, market, market squares and open spaces in the vicinity. The sharing of these common areas involve physical as well as social and symbolic activities and their management, both in everyday life and for special occasions. Through the use of these spaces and process underlying it, ethnic and inter-ethnic expressions, interactions and relationships are constantly evolved to give the "place", that is, Marine Parade estate, the content and meaning of its multi-ethnic character over its 15 years of existence. Its historical background and ethnic mix, in addition to its shared residential features give the estate its significance as an example of the continuous construction of a multi-ethnic community in Singapore.

Malay Special Occasions: Weddings and Hari Raya Puasa

Among all the special occasions which take place in Marine Parade estate, the Malay wedding occurs most frequently, usually on weekends. While a number of the ceremonies for the wedding take place within the flat, others are held in full public view in the void deck space rented from the HDB. These include the *rewang* (wedding preparations such as food and tables), *kenduri* (feast), music accompaniment, *kompang* (hand-held drums) group performance, *silat* (martial arts) display and the *bersanding* (blessings) ceremony.

Hari Raya Puasa[7] celebrates the end of Ramadan, a month of forbearance, endurance and self-denial through abstinence from world pleasures and fasting. One outstanding feature during Ramadan is the holding of evening prayers among Muslims in some void decks. Organised by a Muslim block or neighbourhood committee, the mutual help and solidarity of evening prayers are expressed not only in joint prayers itself but also in its organisation and space preparations such as lighting, laying of prayer mats, cleaning and the sharing of food. This community spirit is also expressed in the morning celebration prayers held in a huge marquee specially set up in the basketball court.[8] Another outstanding feature is the "light-up" in which thousands of colourful lights illuminate the entrances, windows and corridors of virtually every Muslim home, collectively turning the whole estate into a twinkling maze. It is common knowledge that from the lights, one can tell how many Malays live in a community and which blocks they occupy.

These special occasions contribute to the common perception of Malays being highly visible. Via this visibility, they voice several meanings about Malays, both to Malays themselves and to others. The strong sense of local and ethnic identity through these highly visible occasions characterise Malay residents in general. Although holding weddings within home grounds may be rooted in past village forms which no longer exist, the idea itself — that the home is the appropriate site for the rite of marriage remains largely intact. Thus, far from disappearing, the public displays of a Malay wedding form an integral part of the one's auspicious activities and hence, also of the local community's scenery. A Chinese resident, Tee, observes, "You can say that it is a typical feature of life here already". At the same time, being Malay events, they involve members of the Malay ethnic community as kin, friends, neighbours, co-residents and co-religionists,[9] bringing together different levels of meanings — individual, family and community, and religious as well as cultural aspects of being Malay.

Being symbols of ethnic community, the occasions "speak" to the community's members and to non-Malay members, that is, they take on an added dimension — the display of ethnicity to others. A resident, Ros, put it simply and succinctly, "We do all these things so that people know about us Malays". While this display to others is to be seen in the wider perspective involving the position of the Malay minority within the social, economic and political contexts of Singapore, our focus here is on the inter-ethnic relevance of the occasions at the local level.

First, it should be re-emphasised that, because of the public ceremonials, the Malay celebrations form part of the familiar scenes through which non-Malay residents build up a sense of their living environment. Apart from everyday public and personal relations, the wedding also forms the single most outstanding representation of Malayness to the non-Malays, and shape their perceptions of the Malay community. These are illustrated in the following observations:

> See the tent, pots and cooking only, you know it is a Malay wedding. And when it's Hari Raya you see all the lights. The lights make our place really nice. (*Sim*)

> There is at least one on most weekends, and it is usually on Sundays so we can hardly have any rest. They are very noisy. ... I think that for such a special

> occasion, they should hold it in the community hall or hotel, why in the void deck? The void deck is for residents to sit or walk past. What kind of wedding is it anyway with people looking and walking past? (*James*)

> I have been to a Malay wedding ... I tell you, it is a waste of time and money. I told my husband our $20 gift wasted. You know what they serve? Seven of us sitting around one section, but they just put four simple dishes for us to share. I don't think they are even serious about their weddings. So we never went for their weddings after that even though we were invited. (*Mrs Meyer*)

> I think they don't know how to eat well. They prefer to spend their money on display and jewellery. You look at the number of times they change their clothes and how nicely they do up the bridal chamber ... and they are not serious people. You look at their weddings, it is not like the Chinese or Indian wedding, there is no ceremony. ... The bride and groom are not there to meet and greet you. ... And the food is quite simple and quite little. (*Karen*)

> When I see the Malay wedding, I think how nice and simple and how they do things together. Not like we Chinese, too complicated, want to have face, make it so grand. So wasteful. (*Mrs Wong*)

> I don't have Malay friends but I notice their weddings ... everybody comes to help ... it is full of traditions, not like us Chinese. (*Julie*)

> I have heard other people say it is so noisy but I don't hear it. ... Live in this type of place, must get used to it. The wedding lasts only for two days, our Chinese funeral also lasts for about two days, about the same. (*Ah Sin*)

In the cross-cultural contact involved, comparisons are made which reveal different expectations of the rules, formalities and general conduct surrounding the rites as in relation to food, treatment of wedding guests and venues. Varying opinions on the appropriateness of venue as well as notions of private and public reflect different views of spatial and symbolic expressions of a wedding occasion. A Malay resident, Daud, explains:

> Chinese wedding in the house very simple but dinner at hotel very grand, for us, in the house very grand. It is a family occasion so we want people to come to our house. That is only a hotel, why make it so grand?

Behavioural awkwardness can also arise from these cultural differences such as where guests of different genders may sit. Boundaries are drawn as to which rituals guests can participate in or not. Judgements made in cultural comparisons become more contentious when they result in negative ethnic stereotyping.

On the other hand, some Malay residents refer to what they see as unfair treatment by residents and the HDB officials regarding Chinese and Malay special occasions.

> What I think is unfair is that when the Malays burn the turf when they do cooking, they get their deposit penalised but when the Chinese burn the grass when they pray, they don't get fined. (*Zul*)

Under the terms of renting space from the HDB, the user's deposit is penalised if there is any damage to property. In contrast, it is difficult to catch "in the act", the individual Chinese resident who damages the grass turf when burning ceremonial joss papers. Large-scale burning of these papers during the Seventh Month Ghost Festival is so widespread in the predominantly Chinese estates that it becomes virtually impossible to control.

Similarly, the regulations regarding noise can assume ethnic overtones in their interpretations. Popular bands are not permitted by the HDB on grounds of their noise level although stage shows featuring these bands during the Seventh Month Festival are included in the official permit. Said the resident, Daud:

> Not so fair, they complain we make noise when we have weddings but we never complain when they make

> noise at funerals or ghost festivals. It is a happy occasion, sure make noise.

While negative responses to Malay weddings may exist, what is significant here is that tolerance and acceptance of each other's cultural practices generally prevail. This tolerance and acceptance as the predominant manner of accommodating cultural differences can be further understood and better appreciated in the context of the other major set of special occasions which the residents also encounter — the Chinese special events.

Chinese Special Occasions: The Seventh Month Ghost Festival, Funerals and Pavement Prayers

The Chinese Seventh Month Ghost Festival is the largest-scale Chinese festival held in the locality. There are varying versions of the celebration and details in different locations but all share the fundamental belief in the appeasement of gods, ancestors and ghosts as well as the attainment of practical goals. In the Marine Parade estate, individual households mark the occasion with offerings in and around their homes and at least four community-based celebrations are held by the Chinese business community and residents.

The festival is celebrated in groups and each group event is organised by a working committee. Each event is held in a huge tent set up in either a carpark or market square rented from the HDB. Festivities include prayers and offerings, a dinner for several hundred persons, an auction to raise funds and a stage show, all of which, provide much atmosphere, with the loud auction bids and stage music reverberating through the entire neighbourhood till midnight.[10]

The Seventh Month Ghost Festival expresses a collective sense of belonging and identity for the local and ethnic Chinese communities. The first is derived from the event's initiation, organisation and participation by local residents as well as the general desire for spiritual protection and well-being. At the same time, it is an ethnic occasion that draws the Chinese together. The festival also reflects other Chinese elements —

economic culture, social sentiment and immigrant history — all of which, lend the event its Chinese character, cultural content and communality in their variety of meanings.

Like the Malay wedding, the Chinese funeral is the most common individual event that is identified with Chinese culture. Also held in the void deck, it is characterised by the high visibility and loudness of the wake which usually lasts two to four days. Wreaths and cloth-blanket banners are displayed while the rented void deck space accommodates coffin, altar and seats for guests. Nightly prayers by priests accompanied by loud music, chanting and sometimes, opera and *mahjong* games form part of the wake.

In addition to the above occasions, pavement prayers occur on many other occasions, both personal such as to appease a spirit and cultural-religious as for Chinese Lunar New Year. Public areas such as the aprons of apartment blocks, grass areas and around trees are the main sites for these prayers and offerings. Temporary or even permanent shrines have been set up.

While the occasions are significant to the Chinese community as local and ethnic events, they also affect relations between Chinese and non-Chinese residents. Their inter-ethnic relevance arises from their frequent occurrence in common spaces and in particular, their material and symbolic expressions which residents unavoidably encounter. Residents' responses show the means through which they resolve and accommodate cultural-religious differences.

> At first astonished, felt angry. So much noise, cannot sleep. ... next day got to work. .... Also so much ash. But after a while, accept it. Just shut the window and accept it. Must accept, Chinese got their own way. Living together, give and take. (*Yusuf*)

> We have our customs, they have their customs. Sometimes they pray the ash flies inside the house but that is only once in a while, we just clean it up. (*Haja*)

> I avoid the burning and offering if I see it, I take a different path. ... My Chinese friends also tell me not to step on it. (*Neeraj*)

> The Hokkien and Teochew [Chinese dialect groups] ... also got *lok tang* (trance), we Malays say *menurun*. Actually, Chinese, Malays and Indians also have *menurun* but for me ... that is *syaitan* [satanic]. Used to it already. ... Sometimes they like to put things under the tree. ... I just tell the *datuk* [shaman believed to inhabit the tree] I am cleaning the place for you. ... I tell the *datuk* by thinking it in my head. (Sweeper)

> I know about the 7th moon, trance ... I go and take a look. Some of my Christian friends said I should not. ... From what I know, usually it is because something happened or someone is sick, so they ask the medium to do something to help them. ... it is a form of praying. They have their own way of reaching God ... (*Mrs Roberts*)

> I don't mind the burning. I just make sure I don't put my foot there, something can happen to the leg. I believe, better to believe, it ever happened to my son. He was young, he didn't know so he must have walked on it or kicked it. ... He got fever for many days. We took him to hospital but no use, he didn't get well. Then we took him to a *sami* [priest] in an Indian temple, he said my son stepped on something. Then he prayed and gave my son something to drink and after that, he was okay. So when they burn, I avoid. (*Mrs Singh*)

Barring the initial encounters, most non-Chinese residents regard the above practices as familiar and take for granted, aspects of their living environment. The spillover of noise, ash and offerings are regarded at different and interchangeable levels as physical matter and as symbolic meanings. Physical spillover is dealt with as physical matter out of place — by cleaning up the ash or shutting out the noise. Failing this, one simply bears with it. Unhappiness is seldom openly and directly expressed. On the symbolic level, spillover is regarded as the embodiment of culture and religion for which, respect is due and right of practice accorded. They are unquestionable aspects of the culture of ethnic others. By extension, this view is also the basis for the overall approach of mutual "non-disturbance" and peaceful coexistence of different religions.[11]

Tolerance and acceptance of Chinese religious activities are also motivated by practical consideration and responses required of a "no-choice" situation. This is expressed in "live in this type of place" or "live in Singapore" which may be taken to mean that there are no alternatives to public housing or a living environment shared with ethnic others and a dominant Chinese population. Hence, the responses of "cannot do anything", "give and take", "happens only once a while" and the "equal balance" of occurrences and spillover from each other's occasions. Inconvenience and irritation caused are kept to the self. Sometimes, tolerance may be stretched to the utmost limit. An unmanageable situation can then be turned over to the HDB for mediation since the consequences of direct negotiation or confrontation on neighbouring ties may be too costly to bear.

The HDB's Management of Ethnic Differences

When the study discussed in this paper was being conducted, the implementation of public housing policies and rules and the everyday management of the estate was still being managed by the HDB staff at the Marine Parade estate Area Office. Since then, there have been major organisational changes which will be discussed in the last section of this paper. At the area office, the staff of clerks, housing and maintenance inspectors (HMIs), parking attendants and cleaners were headed by estate officers (EOs), chief of whom was the head of the area office who would be directly responsible to the HDB headquarters.[12] A number of the Marine Parade area office staff — one EO and several HMIs were interviewed.[13]

The HMI is in most direct touch with the residents. He is responsible for seven to nine blocks of flats, depending on the number of units in each block. This staff member sees to tenancy issues, enquiries, applications and complaints. He or she is in direct contact with the residents and other relevant parties if needs be, and makes a daily round of inspection in the area under his or her charge to ensure that "everything is in order". The HMI works under an EO to whom, problematic matters can be referred to and from whom guidelines and orders are received. In short, the HMI is the local "investigator", "problem buster" and "conveyor" of various matters concerning the estate and its residents. It is, therefore, especially through the HMI and to a lesser extent, the EO, that the HDB's management of ethnic differences at the local level is best understood.

Handling of the special occasions celebrated by different ethnic groups and other cultural-religious expression form a major aspect of the HMI's and EO's work[14] particularly in the early years of the community's formation. The HMI's and EO's involvement usually arises from confidential complaints lodged by residents individually. These residents do not want to risk a direct approach in confronting their neighbours or other residents.[15]

The local HDB officials' sensitivity and management of these and other matters related to ethnicity require a complex combination of the HDB rules and provisions on the one hand, as well as personal knowledge and skills on the other. While such sensibilities and sensitivities are acquired in part through formal training, they are, for the most part, largely developed through direct work experience.

According to HDB provisions, the area office allocates rental space for special occasions through a booking system on a first-come-first-served basis. For a fee, a permit is given for the use of a space for the duration of a special event. The use of open space further away from residential blocks is encouraged to reduce disturbance to other residents. The area office also helps to reduce the irritation from dirt and flying ashes by providing bins for the burning of ceremonial paper offerings, particularly, during major Chinese festivals. Pamphlets are distributed and posters mounted to encourage the use of these bins. The area office makes sure that the bins are of the right colour (red), and draws up a roster for clearing them. Such provisions are the limits of the HDB's role. The rest is largely left to the residents' own devices. Here, the local officials share the same view as most residents regarding mutual tolerance and accommodation of each other's practices:

> Fewer complaints now that they are given designated, approved site; try to be far away from residents. Before, they used to hold them anywhere. ... But Malays understand Ghost Festival now, so can tolerate .... Also, they have permission to use the void-deck for

> their function, so what is there to complain about? (HMI 1)

> Seventh Month celebration is once a year, should be alright. .... Live in HDB, have to tolerate this sort of thing, given and take. Complaints of noise, not so much now. ... Malay wedding, that is not noisy. ... pop band we do not allow. Cassette player yes, ... that is in the permit. They do it illegally by themselves. But if nobody complains, then we don't do anything. We don't go and check on them unless there is a complaint. (HMI 2)

> The damage done to turf is made by both sides, although usually more with Chinese. But in Block 54, the same void deck is used very frequently for Malay weddings because the adjacent block with shop units on the ground floor has no void deck. Permanent patch of bare ground there, no turf because of the frequent cooking. Yet the Chinese family living directly above this patch has never complained of the noise throughout the night or the cooking smell or dirt from the firewood. So you can see that the Chinese also tolerate. Everybody has to tolerate and accept. (EO)

The intervention of the local officials is, therefore, situational. Underlying each case, however, is the careful manipulation of ethnicity — either as a resource to be exploited or as a factor to be played down. Such manipulation in management arises from a high degree of awareness and sensitivity about the implications of ethnic differences accumulated in the course of their work, as well as from their own ethnic backgrounds.

The most common form of the use of ethnicity as a resource is in the form of cultural and ethnographic knowledge of the residents' backgrounds and practices. Applied in explanation and accounting to residents who complain, this body of knowledge forms a useful resource and tool not only in dealing with the complaint *per se* but for furthering inter-ethnic understanding. Three cases reported are as follows:

> Complaints about Malay weddings is usually about cooking smoke and heat, noise. .... Malays like to cook

> their own food! Also, some people may not be able to afford to cater. So I explain this to the complainant. (HMI1)

> ... this Eurasian woman, she complained to me about noise so I quickly rush over. She asked me why the Chinese funeral cannot be held in another void deck, always in the void deck of her block. If she knows the custom of the Chinese, then she will know why they will never go to another void deck, but she didn't know, because she is Eurasian, not Chinese. So I explain to her ... (HMI 3).

> These people who pray at the roadside usually have problems. ... I know all this because when I was young I followed my parents to the temple and they explained this to me. ... And once you understand, then you won't mind, ... even if you don't believe yourself. (HMI 2)

Such knowledge adds sensitivity to the officials' perspective and management of a problem as with an HMI's response to a complaint from both Chinese and Malay residents about noise from a medium's apartment, "I advise about the noise, keep it low and not late when conducting prayer session. You can't say don't do it, that is his livelihood".

The official's own ethnicity is carefully managed, either in positive use or played down, depending on the situation. The main aim is to maintain impartiality. This is illustrated in the following responses:

> We make sure that we are acting as an officer from the HDB and we listen to them as HDB officers, not as a Chinese or Malay. ... you don't take sides ... (HMI 2)

> People complain because someone uses his residence as a temple. Then we send a Chinese guy to deal with him. If you send a Malay guy ... there may be misunderstanding ... (HMI 3).

In the days before the first-come-first-served principle was established, there were conflicts in allocating the use of void deck space. The simultaneous demand and use of the same void deck for a Malay wedding and a Chinese funeral wake is a poignant illustration of such a conflict. The EO explained that in the past, the two sides just ignored each other and used different ends of the void deck space. At times, there have been negotiations, either initiated by HDB or among the parties themselves. Now, the Chinese funeral parlours first check with the HDB office about whether the space has already been booked for a wedding or other functions.

Both positive and negative ethnic stereotypes are also managed; the first as a resource and the latter as a factor to be handled sensitively. The EO explains this in the following manner:

> Apology usually makes the Malays satisfied, their concept of forgiveness very strong, definitely a plus point when working with Malays. Not so with Chinese and Indians, they are more vengeful. Malays are tolerant, accommodating and understanding about culture. Even though they may not agree with your cultural practice in private, they do not criticise others, so long as others do not criticise them and "touch" their religion. But you must be careful with Malays because they are not like Chinese. They still think of their own kind, they will side with their own kind if anything happens. ... Malays still have this type of thinking. For Chinese, this kind of thing not so much. But it depends on education. ... Both sides have their mentality — majority mentality and minority mentality, that is why you have to be careful how you handle a situation. (EO)

The HDB rules provide it with extensive powers of control and punishment, ranging from fines to eviction for violations of lease/tenancy agreements. But, in the reality of local everyday life management, it is through a combination of punishment, provision and personal persuasion that its officials handle ethnic-related matters. In a difficult situation, the approach is to "... let the matter mellow first. We would rather pay the price for being over cautious than not cautious. ... That is why you haven't seen a riot yet. Better let matter mellow first. Yet, have to remove each last straw, so that we can have room to manoeuvre later on, when other issues arise", according to the EO. The high sensitivity to and awareness of ethnic issues is sharpened by the recognition that the construction of a multi-ethnic community acceptable to all is a continuously evolving task and that the seemingly mundane processes of daily negotiations and management involved are important aspects of this community construction.

A cultural and symbolism map can be drawn based on the special and public nature of ethnic occasions to provide visual feasts of the ethnic diversity of the local community. While unavoidable, such multi-ethnicity also provides the basis for defining the residents' sense of familiarity with their environment and place identity. At the same time, much colour and vitality are infused in an otherwise routine everyday life and highly standardised environment (see Chapter 7 on the same subject). Being expressive and symbolic discourses of the social relationships surrounding them, special occasions involve two dimensions — ethnic community and inter-ethnic relations. For members of a similar ethnic group, the occasions affirm and reinforce ethnic community boundaries in the multi-ethnic setting. Ethnic community is strengthened and given meaning and identity. In addition, the events also serve as a means for inter-ethnic knowledge, communication and comparison at the local level.

Malay and Chinese ethnic occasions form part of the repertoire of devices which define what is Malay or Chinese and they are an important part of socialisation and ethnographic knowledge about others. Strong boundaries, however, are maintained and seldom crossed except for rare occasions. The ethnographic knowledge which exists concerns more forms than the meanings. Comparisons and judgements reached on the basis of such knowledge have shown ethnocentrism, cultural ignorance and deprecation by some. But this should be balanced against the tolerance, acceptance and appreciation exhibited by others.

The obvious and taken-for-granted peaceful coexistence of different religions is significant especially in the light of ethnic-religious conflicts elsewhere. Behind it are complex practical and symbolic means of negotiating, coping and resolving religious expressions considered to be physically or ritually polluting and dangerous. Residents cope through avoidance, faith in one's own religion, tolerance, understanding, "let-it-be" attitude and turning to an external authority where things become unmanageable. Fundamental religious differences, it must be noted, especially between monotheistic and polytheistic faiths, may be discussed but seldom challenged inter-ethnically. Where religion is concerned, the residents emphasise differences but also the rights to such differences. Two residents commented:

> You have your own and I have my own. So long as don't disturb each other, there is no problem, respect each other's religion. (*Rosita*)

> All want to live in harmony, why clash? Only religion different. In the end everyone also same God, only way of praying different. (*Yusuf*)

Only in times when the residents are unable to cope or risk direct negotiations with their neighbours of different ethnic backgrounds do they turn to official management. Use of void deck space occurs through a booking system which legitimises the use of the area for the duration of an ethnic occasion and there is no monopoly of the space.

Implications of Recent Policy Changes

Day-to-day living at the local level is affected by rapid economic, social and political changes in the wider society. Changes in official or state policies and management of ethnic-related issues in public housing estate form part of wider social changes. They have direct impact on the continuous process in construction of the multi-ethnic community.

Two major policy changes affecting Marine Parade estate and other public housing areas were the introduction of Town Councils in 1989 and an ethnic quota for housing allocation by estate. Town Council effectively replaces the area office in several functions of maintenance and management of common property (see Chapter 3).[16] This policy-change appears to be in line with a general trend of allowing the public more direct participation in decision-making and has been aimed at encouraging public housing residents to be involved with the management of their estates.

Town Councils, in taking over most of the area office's tasks must minimally be able to continue the work of managing ethnic differences and maintain the general goodwill and tolerance in existence. Some continuity is ensured with some of the former area office staff opting to join the Town Council. In addition, direct input from the residents potentially offers opportunities for promoting ethnic interaction that are creatively conceived and conducted. Town Councils can learn from the HDB area office, both in the management of ethnic differences and the further provision and improvement of estates facilities.

The ethnic quota sets limits on the proportions of residents of Chinese, Malay, Indian and "Other" ethnic groups in every public housing apartment block and neighbourhood. The aim is to check the formation of "racial" enclaves that had been growing in a number of estates. The resale market in public housing apartments had made possible the concentrations of ethnic groups that had, during the allocation of new flats, been spatially dispersed in ethnically mixed estates. Ethnic quotas plug the loophole that had therefore existed.

The study discussed in this paper has shown that the HDB form of shared-living evolved so far is acceptable to most, and that, the residents are able to arrive at some major modes of accommodation over ethnic differences by themselves, independent of official intervention except in some specific areas. While such accommodation derives in part from a no-choice situation, it still does not detract from the study's main findings.

Marine Parade has a proportion of Malay residents that exceeds the proportion of Malays in the national ethnic profile. The formation of concentrations of ethnic minority groups in certain estates must be understood in multi-dimensional perspective. This study emphasises that an ethnic concentration is essential for the existence and sense of well-being of members of an ethnic community and particularly, that of a minority within the larger multi-ethnic environment as well as being compatible with it. Other studies show that family and extended kin are the main sources of social activities and support such as in childcare (Wong et al. 1985). These aspects of an ethnic community have many necessary and positive features which should not be seen as an unwillingness to mix with others or in conflict with ethnic integration. If enclaves are seen as the traditional preference to live amongst one's own kind, then it should be addressed as the reflection of lack of ethnic integration in the larger society.

This study shows no serious problems of ethnic integration within the residential community itself. Reasons behind the trend towards the concentration of ethnic groups have to be found in other areas of social life such as economic status and educational achievements. Fixing residential quotas may, at best, only contain the size of ethnic enclaves; it does not increase integration in either the residential community or the wider society.

More needs to be known about how strong attitudinal preferences are for living with one's own ethnic group and unwillingness to mix compared to the practical reasons of family and social support. Both minority and majority mentalities should be further understood. Policy-makers should consider, on a continual basis, the identification and building on other non-ethnic factors to increase the range of opportunities for inter-ethnic contact, reduce exclusively ethnic associations and diminish dependence on ethnic-based resources such as facilities that are ethnically exclusionary.

The quota policy, to ensure ethnic harmony in the long term, is well-meaning but appears out of touch with the complex historical and social reasons for ethnic concentration. This policy appears to have exceeded the boundaries of acceptable state intervention and may have negative impact on the generally good record of the HDB and the government as well as on ethnic relations in general. While interviewees in this study do not deny the need to ensure ethnic harmony and integration in public housing communities, they question the policy's desirability on the following grounds:

- it is discriminatory against minorities since the Chinese are a majority in every estate;

- the individual loses the right to choice of residential location and

- the policy ultimately reinforces negative thinking along racial lines as it introduces a conscious criterion of "race" in the choice of residence.

Résumé

In the final analysis, what is ethnic integration and how it is to be achieved is a question and task to be constantly faced. In the context of nation-building, social engineering of ethnic integration by the state requires careful and sensitive management. Too much and it might backfire, too little and it has no effect. In this, the use of national population proportions as a basis or guide for decisions about ethnic representation should not be indiscriminate or become an ideological trap.

Generally, the local officials' awareness, discretion and flexibility on the one hand, and professionalism, caution and fairness on the other, is illustrative of the sensitivities and sensibilities necessary for the official management of ethnic differences. They demonstrate that official intervention has a necessary and useful role in the continuous construction of the multi-ethnic community through the mediation and negotiation in specific situations of inter-ethnic disputes as well as in identifying problems and promoting cross-cultural understanding through direct observation, mediation and explanation. In doing so, they also gain clout and credit for HDB's policy of ethnic integration through shared residential living. At the same time, more space must be allowed for the residents' spontaneous, independent and gradual processes of learning to live in a multi-ethnic environment.

ENDNOTES

1 While the residents also participate in local life through group activity and organisations such as the Residents' Committees, block committees and the community centres, this study focuses on the individual resident and the individual level of encountering ethnic differences.

2 Everyday life is the other major social arena of ethnic interaction in the residential community, and includes modes of dressing, eating, shopping, chatting and playing. For a detailed discussion, see my monograph *Meanings of Multi-ethnicity — A Case Study of Ethnicity and Ethnic Relations in Singapore (1995).*

3 Most discussions of rituals and festivals assume a homogeneous society in which its members share or are expected to conform to their belief systems.

4 The original residents still comprise the majority of Marine Parade's population at the time of the study. However, between 1985 and 1988, there was active movement of residents into, out of and within the estate. Many young residents moved out to set up new households in other estates and new residents moved in to replace those who had left during this period. Other residents also moved from rental to purchase flats or from smaller to bigger purchase flats within Marine Parade. These movements were possible through the buying and selling of flats in a highly active resale market, mainly three- and four-room flat types. Indeed, Marine Parade, with its additional features and advantages of small size, good facilities, proximity to other areas, sea, park, was among the top ten most active housing estates in the overall HDB resale market. All three movements of residents cut across diffferent ethnic backgrounds although exact figures are not available. However, the general trend among the new residents is that there were more Malays among those who moved into three- and four-room flats and more Chinese among those moved into four- and five-room flats. There were also more Malays among those who moved from smaller to bigger flats within the estate. The trend for those who left the estate is less clear, other than that it cuts across all ethnic backgrounds, with those acquiring private housing being mostly Chinese.

5 Estimate given by former HDB Area Office.

6 The void deck is the covered space below the flats on the ground floor of the block.

7 *Hari Raya Haji* is the other major event which draws the ethnic community together. It is a reminder to Muslims to sacrifice their bounty in the name of Allah and is marked by prayers and symbolic sacrifice with the slaughter of sheep at mosques. In Marine Parade, it is expressed by mass prayers in a marquee and the wearing of ethnic or religious attire by the young and old.

8 This arrangement is to avoid overcrowded mosques and for those who wish to pray within their residential community.

9 While Arabs, Indian Muslims and other Muslims also particpate in mass prayers on religious occasions, the vast majority of co-worshippers are Malays.

10 Foods are also distributed to members as part of their sharing in the celebration, and are of strong practical and symbolic significance in Chinese culture.

11 In varying degrees, the animistic basis of spirits persists and is blended with the different religions at the folk level of interpretation. This often finds expression in shrines (*shen*, *datuk*, spirit or saint) common to Hinduism, Chinese religion, Malay animism and Catholicism. Some Chinese tree shrines incorporate Hindu and Malay animistic beliefs in spirits and holy places *(keramat)* which the Chinese call *latuk kung*. The term *latuk* is a Chinese version of the Malay word *datuk* (grandfather) which, in this context, means a shaman (tree spirit).

12 In the hierarchy of job mobility within the HDB, the HMI can attain the level of Higher HMI (HHMI) and EO, while the EO can attain different executive levels of rank and promotion.

13 The EO and HMIs cited here are of Chinese and Malay ethnic backgrounds. The EO has been with Marine Parade Area Office (AO) for two and a half years and was with another Area Office previously; HMI1 has been with Marine Parade AO for eight years, ever since he first joined with HDB; HMI2 has worked in two other area offices near Marine Parade before joining its AO seven years ago; and HMI3 who has been a MHI for five years has just been transferred to Marine Parade from another estate office.

14 Another major area of social problems pertain to economic hardship cases in which the residents are unable to pay their rental or purchase instalments.

15 HMIs observe the rule not to divulge names of complainants to others, and in any case, most complainants request confidentiality on the grounds that their relations with neighbours may, otherwise, be adversely affected. One HMI told me that when he makes home visit while investigating complaints by one party about another, he does not meet one immediately after the other, just in case, he is being observed by either party.

16 The HDB continues to be responsible for administering tenancy and lease covenants, parking and repairs to internal areas of rental properties.

REFERENCES

Cohen, A.P. (1985) *The Symbolic Construction of Community* (London: Ellis Horwood, Chichester and Tavistock Publications).

Douglas, M. (1985) (reprint) *Purity and Danger — An Analysis of the Concepts of Pollution and Taboo* (London: Ark Paperbacks).

Frankenberg, R. (1978) (reprint) "British Community Studies — Problems of Synthesis", in M. Banton (ed.) *The Social Anthropology of Complex Societies*, ASA Monographs 4, (London: Tavistock Publications).

Lai, A.E. (1995) *Meanings of Multi-ethnicity — A Case Study of Ethnicity and Ethnic Relations in Singapore* (Kuala Lumpur: Oxford University Press).

Lewis, G. (1980) *Day of Shining Red — An Essay on Understanding Ritual* (Cambridge: Cambridge Unviersity Press).

Wong, A.K., Tan, T.W., Chew, S.B. and Lai, A.E. (1985) "Family Lifestyles Among HDB Residents", in A.K. Wong and S.H.K. Yeh (eds.) *Housing a Nation — 25 Years of Public Housing in Singapore*, Maruzen Asia for Housing and Development Board, (Singapore).

10

Singapore's Garden City: Translating Environmental Possibilism

Victor Savage

Introduction

The United Nations 1992 report on Environment and Development cited Singapore's "green" policies and its integration of environmental concerns into development policies as a model for developing countries (*Straits Times*, 27 May 1992, p. 25). In a recent green audit conducted by renowned conservationist, David Bellamy, Singapore scored seven out of ten points in environmental management. According to the botanist and globe trotting ecologist, "Singapore comes out very, very well in the assessment, and is a role model for other cities in sustainable development" (*Straits Times*, 4 May 1993, p. 24).

The high rating by the United Nations and Professor Bellamy on Singapore's achievements in environmental management begs the question of how Singapore has been able to balance high economic growth rates and maintain a clean-and-green city. This paper argues that Singapore's success as an environmentally friendly city is a product of the political leadership's pragmatic environmental ideology that is hinged on possibilistic and anthropocentric human-nature relationships.

I have used the concept of environmental possibilism in contrast to environmental determinism in the sense that human beings, societies and civilisations are not influenced by environment. Possibilism is a school of thought which upholds the view that humans, by virtue of their culture, intelligence, pragmatic philosophy, technology, organisational abilities, among other things, are active participants in human-nature relations (Savage 1992, p. 189). The possibilists thus see environment as setting opportunities and possibilities in which man is the final arbiter and judge of their use. Both colonialists and Marxists have used the possibilistic credo as governing ideologies (Savage 1992, p. 189).

Since the People's Action Party (PAP), under the leadership of the former Prime Minister and now, Senior Minister, Mr Lee Kuan Yew, came into power in 1959, there has been a dramatic and drastic change in the Singapore landscape. Over the last 30 years, ten per cent of Singapore's land area (about 6000 hectares) has been reclaimed. The coastline has doubled in length from 140 to 300 kilometres. Some 86 per cent of the people live in public housing flats and estates. Several new reservoirs have been created, rivers have been canalised and industrial estates developed. Over 2000 kilometres of roads, highways and expressways have been constructed. 50 per cent of the total land area is now classified as built-up. Some five million trees and shrubs have been planted, however, and close to 1800 hectares of manicured parks, gardens and recreational space have been developed.

The Singapore urban experience is, perhaps, the best testimony of a planned "conquest" and modification of landscape. It is also a reflection of Gutkind's (1956, pp. 11–12) "I-it" relationship where organic, symbolic and magical human environment bonds are now replaced by abstract, confident, aggressive, utilitarian and expoitative attitudes towards the environment. In the process, Singaporeans have developed through their science, technology, energy and money, together with huge investments of time, an engineered landscape.

Green Singapore: Ecological Reality, Pragmatic Concerns

The then Prime Minister, Mr Lee Kuan Yew, had as early as 1963, initiated the tree-planting campaign. This marked the beginning of the government's concerted efforts at the greening of Singapore. For 20 years, that is, from 1971 to 1990, on the first Sunday of November every year, Cabinet Ministers and Members of Parliament have recreated the political ritual of tree-planting day, under the spotlight of mass media coverage.

Since 1963, over 730 000 trees have been planted together with three million shrubs. Some 4460 hectares of horticultural estates have been developed. This greening of the city by the government is not the product of demonstrating romantic environmental ideals *per se*. Tree planting and the greening of Singapore in the Singaporean context have always been viewed as a pragmatic necessity for the city-state. At the official level, the rationale for tree planting has been seen as a means of providing shade and shelter, containing flying dust, reducing noise for public housing flat dwellers, adding greenery to a concrete landscape and changing the street-level micro-climate of tropical Singapore. The leadership's obsession with maintaining a clean-and-green city led to the establishment of the Ministry of Environment in 1972. Singapore became one of a handful of countries in the world to establish an Environment Ministry at that time (Savage and Lau 1993, p. 14).

Over the years, the creation of the garden landscape has undergone several changes. Besides the initial greening programme, the authorities have added flowering plants in 1982, fruit trees in 1982 and 1983, plants and trees to enhance bird life in 1983 and 1984 and scented flowers in 1990 in its neighbourhood parks and gardens (Savage 1991, p. 68). The area allocated to parks and gardens covers three per cent (1800 hectares) of the total land area in Singapore. This concern with parks and gardens reflects the government's politically astute realisation that, as Singaporeans move up the economic ladder and demand a better quality of life, they would want to enjoy good recreational grounds and live in pleasant and aesthetically pleasing environments.

Whatever the aesthetic, ornamental and recreational ingredients in the garden city, the greening of Singapore has more direct ecological implications that are often overlooked. The vegetation balance must be maintained because it consumes carbon dioxide emissions and enhances the supply of oxygen. One must not forget that it takes 80 trees per person to maintain a balanced atmosphere. For an urban environment like Singapore, the balance between vegetation and concrete built-up area cannot be overlooked. Governments and the political elite are only too familiar with the problems of air pollution, smog and acid rain that have become critical to human health in cities like Los Angeles, Bangkok, Mexico City, Taipei, Hong Kong, Jakarta and Manila.

In Singapore, the greening campaign and strict anti-pollution laws have an additional pragmatic reality. Water is Singapore's Achilles heel. Singapore relies on Malaysia for 60 per cent of its water supply. This gives the country a precarious, strategic and political problem. No country is quite as vulnerable as Singapore with regard to such a basic life-sustaining commodity. The Singapore leadership's concern with water is reflected in the response to a proposal for a zoo to be built next to Seletar Reservoir. The now Senior Minister, Mr Lee Kuan Yew, noted that his concern was with the danger of the zoo polluting the reservoir, "This [the zoo] is bang on a reservoir for drinking water. ... I was obsessed with water" (*Sunday Times*, 20 July 1993, p. 3). This explains why every river system in Singapore has been created into a reservoir.

Reservoirs are, however, useless without proper catchment areas. Dense vegetation around reservoirs is thus an ecological necessity. The Singapore authorities are certainly mindful of the need to maintain a vegetative cover on the island though there has been some debate about the sort of vegetative cover that needs to be maintained.

In a similar way, the Singapore government must also be concerned about the rampant deforestation taking place in the region. With its dependence not only on Malaysia, but soon on Indonesia for water, the Singapore authorities cannot be blind to the impact of deforestation on water resources. Singaporeans cannot turn a deaf ear to repeated media reports of the environmental problems that many neighbouring countries are facing, as a result of rampant and widespread deforestation such as drought, floods, climatic changes, agricultural disasters and soil erosion.

A most poignant reminder of the vulnerability of water shortage was the recent drought and water shortage that the nearby state of Melaka in Malaysia faced. The reservoir serving the entire state had been depleted of water due, in part to, heavy deforestation in the catchment area. The government's concern about the region's forest is underscored by the Ministry of Environment's pamphlet on "Tropical Rainforest" that concludes with this reminder, "Since tropical rainforests serve an important part in the balance of our ecosystem, we should try to conserve these rainforests as much as possible," (Ministry of the Environment 1991).

Human Environment Relations: Anthropocentrism

To ensure that Singaporeans live in a healthy environment, the government has provided a holistic and integrated approach in its environmental policies. Tree planting alone, cannot stop air pollution. In the Singapore case, the government has introduced bold policies to prevent pollution of its air, land and waters. The control of the car population in Singapore through the novel "certificates of entitlement" (COE) bidding system for vehicles is one example of controlling the car population in Singapore and in turn, reducing air pollution.

Singapore maintains a sophisticated waste control infrastructure. All industrial and 97 per cent of residential premises are served by 2145 kilometres of central sewers and sophisticated waste treatment facilities. The Ministry of Environment operates 200 garbage trucks for the disposal of domestic and industrial trash of the city. The island has three incineration plants that handles urban trash.

"Unhygienic" street vendors have been taken off the streets and centralised as well as licensed to operate in large hawker centres. These centres have been provided to ensure sanitary cooking, clean and proper drainage facilities. Such cleanliness, however, has not come as a way of life. In Singapore, the clean-and-green campaign is augmented by strict regulations, enforcements and penalties by the Ministry of Environment. As an example, there were 516 customer complaints about hawkers in 1988, some 844 summonses issued and 337 warnings to hawkers.

The Singapore official approach to human environment relationships has tended to take a more anthropocentric rather than ecological perspective. In the context of Singapore, as a city, this equation seems to be realistic and pragmatic. Cities are, after all, human constructs and artifacts of human engineering, science and technology. It seems difficult to deal with the city as an ecosystem because urban communities tend to beat to a cultural rather than natural rhythm.

In this context, Commoner's (1971) third law of nature that "nature knows best" does not apply. We are dealing in the city, with a concrete jungle and the only law that can be applied are human laws. The Singapore's official view of the garden city is, thus, an endorsement of the anthropocentric bias on environmental issues. While environmentalists and green academics have been concerned with ecological and human ecosystem paradigms, the Singapore leadership has placed greater emphasis on the human dimensions of environmental issues. Several themes in Singapore's development programme have underscored this anthropocentric bias.

The first is the government's political faith in urban planning. The government's early interest in planning reflected the ruling party's democratic socialistic ideology. Over the years, however, planning in all aspects of Singapore's landscape has been a political, economic and social necessity governed by spatial limitations. Moreover, since the city is a human construct, the officials have placed a deliberate and conscious policy of urban planned development, despite the fact that market factors have had some impact on the location and value of land use. These have already been discussed in Chapter 5.

The government has controlled urban land-use development largely through laws, Planning Acts, zoning systems, development controls. Some 76 per cent of the land is also owned by the state. While earlier urban plans have dealt with public housing, industrial, transportation and military land-use development, more current plans have displayed a greater emphasis and sensitivity to the provision of open space, gardens and parks as well as the preservation of natural habitats. With the current interest in the environment, the Singapore Ministry of the Environment launched its Green Plan in May 1992. The Green Plan hopes to make Singapore a green city by the year 2000. This Plan, once again, places the onus of protecting and conserving the environment on the human dimensions.

What has made both urban and environmental planning important in the Singapore context is the fact that, unlike other countries, Singapore has limited land resources and this has compelled the government to be prudent in land-use development. After considerable debate, the Singapore authorities have tried to create more space not only through land reclamation projects but also by ensuring that the majority of Singaporeans live in high-rise and high-density housing.

In the next 20 to 30 years, the government hopes to increase Singapore's land area by another ten per cent through land reclamation. Even prior to the European Economic Community's proposal of building "Compact Cities" based on urban intensification, energy efficiency and environmental protection (Commission of the European Communities 1990), Singapore's urban development has been based on creating a compact city. High-density and high-rise living have provided the rationale for freeing land for open space and green areas and in turn, allowing the conservation of natural habitats.

Singapore's human-centred orientation is also evident in the government's strict enforcement of environmentally friendly conduct through laws and moral suasion. The government's clean-up and greening campaigns would not have been possible without the plethora of stiff fines imposed for a variety of eco-unfriendly activities — littering, spitting, smoking, killer-littering, not flushing toilets, misuse of dustbins and more recently, sale of chewing gum.

In Singapore's political calculations, human welfare is viewed as an integral component of development. Environmental issues, thus, cannot be separated from the fundamental economic, political and social issues of society. This perspective has both plus and minus scores for Green adherents. On the plus side, the Singapore government's goals include environmental components as part and parcel of development issues. On the other side of the balance sheet, officials see economic, social and political priorities at times outweighing environmental issues, and the former issues thus take precedence and importance. These dual perspectives have become areas of contention between the Singapore Nature Society and the authorities, especially with regard to the conservation of wildlife areas.

The government also sees environmental protection and care as a responsibility of every Singaporean and not something solely in the charge of the public service. As what the former Minister for the Environment, Dr Ahmad Mattar pointed out, the responsibility of looking after the environment is a collective effort that requires each and everyone to do their bit. So the "Clean and Green" week is a campaign effort aimed at inculcating in Singaporeans a greater sense of responsibility for the environment they live in. The strict enforcement of environmental laws must be seen as part of the effort to uphold communitarian values within the context of environmental responsibility. As with other laws, the authorities see the importance of enforcing strict environmental laws and fines because the community's welfare must prevail over individual misbehaviour.

The issue here is that the case for the environment cannot be left to individual whims and fancies. This is reflected in the ban against the sales of chewing gum. An executive of the Consumer Association of Singapore (CASE), Mr Albert Tan commented, "Chewing gum has given rise to litter and cleaning problems — problems that would take the money of all taxpayers, gum chewers or otherwise to resolve. In this context, the cost of rectifying the problem outweighs the pleasures enjoyed by a few, so CASE supports the ban" (*Straits Times*, 11 January 1992, p. 25).

Singapore's anthropocentric bias in human environment relationships is also evident in the government's heavy emphasis on infrastructural development and organisation. What makes the Singapore urban environmental experience unique is the fact that there has been a great emphasis on the development of a well-built landscape for the health, convenience and well-being of the citizens. In the built environment, such an emphasis has been translated in terms of public housing, hawker centres, satellite towns, good drinking water, modern sanitation and a comprehensive refuse and garbage system. No amount of clean-up and tree planting campaigns would have made Singapore a clean-and-green city if the urban infrastructure was not developed.

The Singapore government's development programme has, to a large extent, been focused on enhancing the quality of life and standard of living of Singaporeans. In this sense, the transformation of Singapore into what appears to be an environmentally friendly city, has a lot to do with the goals and values that the authorities have placed on development. The inter-relationship between development and environmental goals is, therefore, compatible if we are concerned with the quality of life and standard of living of the citizens. What many countries might have taken for granted, is that the Singapore leadership has made basic in its developmental goals such as the provision of clean air, clean water, stable and efficient energy supplies, good housing, healthy environment, modern sanitation and the provision of basic infrastructural amenities.

The Singapore political leadership has shown that, being clean, green, healthy and environmentally friendly is beneficial and contributes to the health and well-being of the people so that they can not only be developmental goals but in a broader context, environmentally friendly ideals. This is, perhaps, the most powerful eco-development message that Singapore has provided for other tropical cities to emulate. When seen in the historical context, what Singapore has achieved over the last 30 years is, certainly, remarkable. For more than a century, colonialism fueled by possibilistic justifications attempted to debunk the general western views of environmental determinism of tropical regions — the view that the tropics were disease-prone, environmentally unhealthy and hence, beyond development and civilisation (Savage 1984, pp. 21–22 and 179–187).

In 30 years, the Singapore government has changed the tuberculosis-ridden and disease-prone slum-and-squatter city into a near-utopian salubrious environment fit for quality living equal to the cities of the developed temperate world. Singapore's garden-city experience has clearly been successful and it is not by accident that Kuala Lumpur and Jakarta have followed closely in developing their own versions of the tropical garden city.

Garden Symbolism: The Engineered Landscape

The garden has, for centuries, been a power symbol of many things. In a young country like Singapore, which celebrated its

25th anniversary of independence from colonial rule in 1988, the garden serves as a powerful national symbol. For Singapore, the garden-city concept has been used not only as an advertisement for tourists but it is becoming increasingly a symbol of identity and pride. To enhance quality living, the government has proposed S$500 million for the development of parks and gardens, over the eight years from 1989 to 1997. This concerted policy to transform the island landscape into a garden city provides a visible demonstration of the quality of life and level of development that Singaporeans have attained.

The garden is a testimony of a people's ethos, cultural attainment, aesthetic taste, political philosophy and man-environment relationship. The Minister for Information and the Arts, Brigadier-General George Yeo, has acknowledged that, to create Chinese or Japanese gardens, "you need a cultural tradition, a certain philosophical approach to life, and above all, a population attuned to the finer aspects of such gardens" (*Straits Times*, 5 June 1993, p. 3). In the Singapore experience, the garden concept underscores the political expression of Singapore's possibilistic political credo. The garden epitomises the human mastery over environment. In the Singapore context, the garden landscape reveals the handiwork and effort of human ingenuity and skills. This is translated as a garden landscape of cropped lawns, pruned bushes, orderly rows of trees, manicured vegetation and regulated nature.

In a garden city, citizens become participants of an engineered landscape. The authorities plan, construct and develop the garden while the citizens are meant to ensure that they maintain parts of this urban garden. To create a garden city of such a scale is no mean feat. This is the archetypal demonstration of the government's planning abilities. The garden, in the Singapore context, is a politically inspired as well as pragmatically managed and "designed" earth landscape. In Singapore, every tree and dustbin has a designated public place, a product of overall design and conscious policy decision-making. The "garden" in the Singapore context underscores urban and environmental planning.

Singapore's environmental planning has been based very much on an integrated and holistic process within the wider governmental master plans. Up to 1992, there was no specific environmental planning in Singapore. Environmental issues were integral parts of the master plans and the concern for the environment was dependent on other land-use demands generated by housing needs, industries, transportation, military usage and water catchment areas. In the frenzied years of urban development (1960–1985), the rapid process of development led to an expansion of the built-up areas. This has regrettably also led to vegetation removal, destruction of wildlife habitats, reclamation of coastal areas, levelling of hills, filling-up of swamp and river areas as well as the canalisation of streams and rivers. With the government's concern about public housing needs and transportation between the various new satellite towns and the cental area of the city, the question of nature conservation was given secondary importance.

Singapore's environmental programme has had more to do with the construction of a landscape moulded and fabricated by human engineering skills than the preservation of natural vegetation. The authorities have argued that, in spite of land scarcity, there are environmental possibilities because emphasis has been placed on good engineering and planning. The importance of engineering in protecting Singapore's water resources has been emphasised (Malone-Lee 1993, p. 9). As an important influence on development, water is often considered as an engineering rather than a planning factor. Singaporeans, therefore, have a good zoological park and clean water although the zoo abuts the Seletar Reservoir. The Senior Minister, Mr Lee Kuan Yew, has underscored the government's possibilistic environmental ideology when he observed that a broader lesson for Singapore is that "Nothing comes naturally .... Behind it is engineering, careful planning" (*Sunday Times*, 20 June 1993, p. 3).

The concern with the creation of a humanly-moulded landscape garden had, by the mid-1980s, become the source of great debate in the mass media. One social worker has commented that "This is a garden city. It is planned such that this blade of grass should grow here, this blade over there. It is nice but it is man-made". This neat and orderly arrangement of plants led one expatriate botanist to lament that, from a biological stand point, Singapore's parks, gardens and road side green corridors are "green deserts where nature enters by stealth, if at all".

Since the 1980s, the new emphasis on environmental conservation has been a result of both international and domestic changes. On the international front, the overwhelming concern with the environment has put both economic and political pressure on many countries. Domestically, the increase in standards of living of Singaporeans has given rise to greater demand for recreational space and the question of open space and wildlife habitats gained greater prominence in official thinking. By mid-1980s, however, the Malayan Nature Society's (now Singapore Nature Society) pro-active role in defence of wildlife habitat in Singapore and the greater media attention on nature and environmental issues (Briffett 1991) led authorities to take note and also take the steps to ensure that wildlife habitats in Singapore are safeguarded. The proportion of land and specific areas to be conserved, however, remains a subject of continuing debate between the government and environmental non-government organisations (NGOs).

Both sides in the debate about nature conservation have compelling arguments. The NGOs see the need to conserve more natural habitats from the ecological perspectives of maintaining ecological diversity, conserving endangered species and upholding Singapore's natural heritage. The official argument for developing some natural habitats is premised upon broader national economic, social, cultural and demographic concerns in land-scarce Singapore. For these reasons, the government has decided to set aside only five per cent or 3130 hectares of land involving some 18 areas as nature reserves. This figure is well below the IUCN's recommended standard of ten per cent for natural areas.

Environmental Possibilism: Science and Technology

> We will build a city of expressways and free-flowing traffic. Of time-saving Mass Rapid Transit, cool and comfortable. Of electronics, telecommunications, and computers.
>
> *Mr Goh Chok Tong, First Deputy Prime Minister (Straits Times, 12 December 1984).*

The national identity of Singapore's garden city has translated Singapore's eco-friendly policies in a unique way. Both the words "garden" and "city" are man-made constructs and hence, the question of environment has to be consciously planned, designed and developed. Unlike other countries that have landscapes of paradise and pristine nature to inspire their citizens and tourists, Singapore's environmental beacon is a product of human vision, creativity, planning and management. This humanly created landscape lies at the heart of the political leadership's credo of an anthropocentric view of human-environment relationship.

Given the small size of Singapore, 624.4 square kilometres or 62 600 hectares, and its resourceless island geography, the leadership has put trust in a possibilistic ideology as the salvation for the city-state. This has been translated into a political mantra of public campaigns, calling for rational, pragmatic and efficient policies, consolidating and rewarding intelligent political leaders and bureaucrats, and defending a meritocratic system.

The translation and success of "possibilism" are visibly manifest in the Singaporean landscape — large reclamation projects, creation of reservoirs from rivers, maintenance of a clean-and-green environment, public housing and new towns, a network of public transportation, highways, expressways and the development of theme parks. While there are no large natural parks offering safari-style adventures, high standards for man-made "natural" creations have been attained — the Jurong Bird Park, the Singapore Zoological Gardens and the Underwater World.

Visitors to Singapore's garden city are often struck by the order, neatness and grid pattern of its landscape. These epitomise the French formal geometric garden tradition and Cartesian science which form the very basis of Singapore's economic and technological progress and the bureaucratisation of human life. Elsewhere, I have argued that the garden-city landscape in Singapore reveals its "political ethos, expressing, perhaps, the social discipline, bureaucratic regimentation, official obedience and the one-party dominant rule" (Savage 1991, p. 74).

The extension of the faith in science and technology has now been translated into part of Singapore's Green Plan. The Ministry of Environment has identified several areas — increasing natural energy efficiency, minimising waste, developing and promoting clean technologies, promoting cleaner modes of transport and replacing ozone-depleting substances. Unlike other countries that have seen environmental programmes evolve in relation to the protection and conservation of environment, the Singapore government's environmental plans are related to reducing pollution and minimising waste and garbage in the city. Despite the fact that the projected amount of waste from domestic and trade sources will reach 1.3 kg per capita per day in the year 2000, the government hopes to bring this waste output down to 0.9 kg per capita per day by the year 2000, a goal targeted in its Green Plan. In 1992, Singapore produced 43 000 tonnes of toxic wastes made up mainly of spent acids, alkalis, spent etchants and solvents from electronic and manufacturing industries.

This great experiment with science or technology on nature is found even in the small sector of the economy that deals with agriculture and fisheries. Given the scarcity of land, Singapore has experimented extensively with breeding of marine life (prawns, fishes and mussles are examples) and hydroponic farming techniques for vegetables. We also have been very successful in plant "hybridsation", that is, cloning and genetic engineering especially with regard to orchids. Because orchid cultivation is a "high-technology" business, the government has concentrated efforts on increasing its cultivation both in terms of expanding farmland — 460 hectares in the next three years — and tripling its exports from S$23 million in 1992 to S$70 million by 1995 (*Straits Times*, 18 June 1993, p. 44). Our national flower, the *Vanda Miss Joaquim*, is an orchid hybrid that symbolises the human experiment with nature.

For the Singapore leadership, natural resource and spatial limitations are setbacks that can be overcome by moving Singapore into the technological age, investing in education and placing faith in science and massive industrialisation. Two reasons were given by the former Minister of Foreign Affairs, Mr Rajaratnam, for the government's faith in industrialisation. The first was that the strongest and most prosperous dynamic societies were those that had advanced technology (Chan and

Haq 1987, p. 255). The second was that based on Singapore's physical limitations, size would become irrelevant once Singapore had developed into a technological and industrial country. This explains why the government has placed great emphasis on telecommunications, bio-technology, computer literacy, high-technology industrialisation and more recently, environmental technology. Singapore's faith in science and technology has been a cornerstone of its possibilistic political ideology and in turn, in the physical transformation of the city-state.

REFERENCES

Briffett, C. (1991) "Proposals for the Conservation of Nature in Urban Singapore", *Environmental Monitoring and Assessment*, Vol. 19, pp. 275–286.

Chan, H.C. and Haq, Obaid ul (eds.) (1987) *S. Rajaratnam, the Prophetic and the Political*, Graham Brash, Singapore, and St Martin's Press, New York.

Commission of the European Communities (1990) *Green Paper on the Urban Environment* (Brussels: EEC).

Commoner, B. (1971) *The Closing Circle* (New York: Alfred Knopf).

Dasmann, R. (1976) "Future Primitive", *The Coevolution Quarterly*, Fall, pp. 26–31.

Gutkind, E.A. (1956) "Our World from the Air — Conflict and Adaptation", in W.L. Thomas, (ed.) *Man's Role in Changing the Face of the Earth* (Chicago: The University of Chicago Press), pp. 1–44.

Lee, S.K. (1992) *More than a Garden City* (Singapore: Parks and Recreation Department).

Ministry of the Environment (1991) "Tropical Rainforests", *Our Environment*, Public Affairs Department, November, No. 16.

Savage, V.R. (1984) *Western Impressions of Nature and Landscape in Southeast Asia* (Singapore: Singapore University Press).

——, (1991) "Singapore Garden City — Reality, Symbol, Ideal", *Solidarity*, Nos. 131–132, pp. 67–75.

——, (1992) "Human-Environment Relations — Singapore's Environmental Ideology", in K.C. Ban, A. Pakir and C.K. Tong, (eds.) *Imagining Singapore* (Singapore: Times Academic Press).

Savage, V.R. and Kong, L. (1993) "Urban Constraints, Political

Imperatives — Environmental 'Design' in Singapore", *Landscape and Urban Planning*, Vol. 25, Nos. 1–2, pp. 37–52.

Savage, V.R. and Lau, S. (1993) "Green Issues — Official Policies and Student Awareness", pp. 13–28 in C. Briffett and L.L. Sim (eds.) in *Proceedings: Environmental Issues in Environment and Conservation* (Singapore: SNP Publishers Pte Ltd).

11

Earthquake Tremors and High-rise Buildings

Kog Yue Choong and T. Balendra

Introduction

There have been many disastrous earthquakes, notably in San Francisco (1906, 8.3 Richter), Tokyo (1923, 8.3 Richter and 100 000 deaths) and more recently, China (1976, 7.2 Richter and 250 000 deaths), Iran (1978, 7.7 Richter and 25 000 deaths), Mexico (1985, 7.8 Richter and 7000 deaths), Armenia (1988, 6.9 Richter and 50 000 deaths), San Francisco (1989, 7.1 Richter and 270 deaths) and Iran (1990, 7.3 Richter and 25 000 deaths). The most powerful earthquake ever recorded reached a magnitude of 9.5 Richter off the Chilean coast. Many earthquakes occur in seismic zones in oceans and remote areas without doing much damage to or loss of human lives. The most disastrous earthquakes occur where areas of high seismicity coincide with large centres of population, notably in Japan, China, California, Central America, west coast of South America, Indonesia, Philippines, Iran and the Mediterranean. Buildings in these areas of high seismicity are required to be designed and constructed to withstand the effects of the earthquake so that damage to buildings and loss of lives can be avoided.

During the Kobe earthquake, which was the worst urban disaster to hit Japan since the great Kwanto earthquake in 1923, a considerable part of the built-up area in the city was destroyed. While the 1923 Kwanto earthquake destroyed wealth, largely in the form of public and private properties which were equivalent to one-third of Japan's gross domestic product (GDP), the cost of the Kobe shock has been estimated at US$200 billion, equal to five per cent of the nation's GDP. Such a figure would be about 20 per cent of the United Kingdom's current GDP (Whittow 1995, p. 81). In the Kobe earthquake, not only were the low-rise older buildings affected but also the modern structures and especially, the elevated highways. The unprecedented increase in the number of recorded disasters in the last 20 years has been highlighted by Ooi and Kwok in their introduction to this book. Indeed, there has been concern that more commitment should be sought from "... national, regional and local governments to disaster mitigation," (Steedman 1995, p. 89).

With world economies expanding into increasingly complex technology and seeking further resources in harsher environments, both people and the built environment face increasing risk from accidents (Degg 1992). The world's mega-cities such as Jakarta and Manila, will have to consider protection for their urban population by promoting designs for good construction techniques and adequate materials.

Over the major part of the globe, no such precautions are needed in the design of buildings. This is the case with Singapore since it is located on a stable part of the Eurasian plate. The nearest seismic faults are more than 400 kilometres away in Sumatra. Nevertheless, since the 19th century, there have been occasional tremors reported to be felt by the residents in some parts of Singapore due to earthquakes in Sumatra. There has even been a reported case of a warehouse roof collapse in 1986, causing injuries to three women (*Straits Times*, 13 August 1986).

It is well known that the effects of earthquake on high-rise buildings are more significant than on low-rise buildings. With rapid economic development and urbanisation and the launch of the public housing programme spearheaded by the Housing and Development Board (HDB) since independence in 1965, high-rise buildings have mushroomed all over Singapore. Therefore, it is not surprising that, since the 1970s, there have been more extensive reports in the local press about the panic caused in some of the public housing estates and high-rise buildings within the Central Business District (CBD) when tremors were felt in Singapore due to earthquakes occurring in Sumatra. Whenever this happens, there will, inevitably, be concern raised by the residents and office workers about the adequacy of high-rise buildings in Singapore to withstand the effects of earthquakes. Typically, the views of the experts and the government authorities would be to assure the residents that there is no cause for alarm.

Generally, the bases for their conclusions are that Singapore is located far away from the epicentres and historically, there has been no reported damage to buildings in Singapore caused by tremors. In addition, they are of the view that complying with the provisions in the code of practice adopted in Singapore for the design of high-rise buildings, will enable the buildings to possess a sufficient level of earthquake resistance.

In 1985, a report on the study of seismic activity over the last 85 years in Malaysia was undertaken by the Meteorological Services Department in the Science, Technology and the Environment Ministry. The report made several recommendations on steps to be taken to mitigate earthquake disasters although, like Singapore, Malaysia is located away from the active earthquake zone in Sumatra and is adopting the same codes of practice as in Singapore. Some of the recommendations are:

a) to set up an interdisciplinary team to evaluate seismic risk and develop an appropriate land-use plan for the whole country, especially over the seismically-active regions;
b) to develop and implement a national building code that takes into account the varying degrees of seismic risk in the country and
c) to design building structures that could withstand earthquake forces in areas of high seismic risk through the use of earthquake-resistant technology being used in countries like Japan and Chile. (*Straits Times*, 16 January 1985).

In fact, according to Professor Chin Fung Kee, a special seismic study carried out in conjunction with the Pulau Pinang bridge project revealed that seismic activity outside Malaysia cannot be ignored in the design of component structural units for the bridge. He further said that the bridge, when ready, will be able to withstand tremors from an earthquake of magnitude 7.5 on the Richter scale with the epicentre at a distance of 300 kilometres. (*Straits Times*, 16 January 1985).

It is apparent from these reports that engineers and scientists in Malaysia are taking the threats of the earthquakes to the built environment very seriously. They, however, realised that there is to date very little work done on seismology and earthquake engineering in Malaysia. Therefore, they recommend that the research and training in seismology and earthquake engineering should be stepped up. There is also an acute need to collect and collate sufficient earthquake activity data to enable a more thorough study to be undertaken. The situation in Singapore pertaining to data collection is similar to that in Malaysia. It is the purpose of this study to examine the effect of earthquake tremors on high-rise buildings in Singapore and to highlight the importance of data collection.

Subsoil Condition

The subsoil condition has a profound influence on the effects of earthquake on high-rise buildings. Therefore, the subsoil condition in Singapore will be examined briefly here. The soil pattern of Singapore is determined by deposition in various periods and has been largely influenced by the effect of the intensive weathering of the land surfaces in a humid and tropical climate. According to the *Geology of the Republic of Singapore* (Public Works Department 1976), the main island of Singapore comprises three north-west to south-east trends tracts of contrasting rock types, as shown in Figure 11.1, with distinctive array of soils associated with each.

The western tract, known as Jurong Formation, consists of inter-bedded and folded sedimentary rocks dominated by shales and quartzite but including sandstone and conglomerate. The soils that occur on these sedimentary rocks are bouldery clay (which consists of a mixture of sandstone or siltstone boulders, cobbles and gravel in a silty clay matrix); clayey silt or reddish brown silty clay residual soil; clayey sand or silty sand residual soil; light grey to white cemented sandy silt or sandy siltstone; and yellow brown, mottled yellow and brown clay of low plasticity.

The central tract comprises igneous rocks dominated by granite but including adamellite, granodiorite, norite and gabbro and is designated as Bukit Timah Granite. The soils that occur on these igneous rocks are bouldery soils with boulders surrounded by residual soil, silty clay or clayey silt, and fine to medium sand.

The eastern segment of Singapore is occupied by alluvial deposits known as Old Alluvium. The occurrence of Old Alluvium is extensive either in outcrop or below layers of the Kallang Formation. It also occurs occasionally, capping low hills of weathered Bukit Timah Granite. The majority of the Old Alluvium is silty to clayey sand. It also consists of sandy and clayey silt. Silty clay and beds of clay of up to 15 metres thick have also been encountered. When fresh, the top of the Old Alluvium is bluish to greenish grey. When weathered, it changes to yellowish, reddish greyish brown. Some of the silty beds encountered at depth are strongly mottled due to weathering.

Throughout and peripheral to these three main tracts of rock types are small areas occupied by recent alluvium, beach ridges and estuaries, each of which gives rise to a distinctive array of soils. The south-east segment of Singapore is occupied by generally unconsolidated recent deposits of variable composition known as Kallang Formation. These soils are peats, peaty and organic clays, silty and clayey sand, sandy silts, silty clays, sandy clays, clayey silts, very soft to soft bluish-grey marine clay, sand with gravels, coral and shells. The maximum thickness of the soft marine clay is of the order of 40 metres. The reclaimed lands at southern part of Singapore, namely, Marina, Marine Parade and Changi are all reclaimed by filling the Kallang Formation.

Macroseismic Study

Collection of past earthquake tremor information up to 1975 is based on volumes III and IV of the series on seismology on Malaysia and Indonesia of the Southeast Asia Association of Seismology and Earthquake Engineering (Leyu et al. 1985, Soetardjo et al. 1985). After 1975, the information is obtained from local press reports. These data are summarised in Table 11.1.

Except for the tremor on 19 August 1977, which was due to an earthquake of 7.7 Richter 2200 kilometres away from Singapore, Table 11.1 summarises all the earthquakes recorded in the existing literature within a distance of 850 kilometres away from Singapore. The distance of 850 kilometres is chosen because no recorded tremors have been reported when the epicentres of the earthquakes are more than 850 kilometres away from Singapore. Where available, the corresponding intensity of the earthquake in Sumatra and the distance of the focus from Singapore are summarised in this table. The distance of the focus in Table 11.1 is estimated from the coordinates of the epicentre; the point at which the body waves reach the surface immediately above the focus. (Note that the focus of the earthquake is the origin of the fracture where the rupture occurs along the fault line or weak point in the rock plates). The earthquake tremor effect is given in terms of the Modified Mercalli Intensity Scale (MM) (see Appendix for details).

It is apparent that the records in Leyu et al. (1985) and Soetardjo et al. (1985) are incomplete because most of the tremors and the corresponding earthquakes are rarely recorded in both volumes at the same time. This indicates clearly a need for more cooperation by the Asean countries in the collation of earthquake data to enable proper studies to be undertaken.

The classification of earthquakes on the basis of depth of focus is: shallow, 0 to 70 kilometres; intermediate, 70 to 300 kilometres; and deep, below 300 kilometres (Smith 1973). It is usually the shallow events which produce the damage at the Earth's surface. The depth of the focus for some of the recent earthquakes after 1962 are summarised in Table 11.1. It is apparent that, out of the ten earthquakes with known depth of focus, only two earthquakes are with intermediate focus and the rest are with shallow focus.

The earthquake intensity after 1965, as reported in Leyu et al. (1985) is in body wave magnitude. These are converted to Richter scale in Table 11.1 using the relationship shown in Figure 11.2. The intensities of the earthquakes in 1952 and 1936, 1935, 1926 as reported in Leyu et al. (1985) are in ROM and PAS respectively. These are not utilised in the subsequent statistical analysis because the conversion of these scales to the Richter scale is not known. Likewise for the earthquake intensities in 1914 and 1909, since it is not clear what scale was adopted then. Furthermore, the earthquake on 19 August 1977 is not considered in the statistical analysis because the focus of the earthquake was not located within Sumatra, unlike the rest of the data.

For earthquakes before 1941, the earthquake intensities are not recorded. Whenever data are available, the earthquake intensities are estimated in Richter scale from the near-field Modified Mercalli intensities reported in Soetardjo et al. (1985), using the relationship shown in Figure 11.2. When the coordinates of the focus of the earthquake are not available, the distance of the focus from Singapore is taken to be the distance from Singapore to the town where damages are recorded, which is usually located within 100 kilometres of the focus. The mean and standard deviations of the earthquake intensities and distances of the focuses from Singapore based on all the data available in Richter scale in Table 11.1 are 5.7–5.9, 0.7–0.8 and 591 kilometres, 114 kilometres respectively. If only the earthquake intensities from 1965 to 1986 are taken, then the corresponding values are 5.8–5.9 in Richter scale with a standard deviation of 0.9 and 567 kilometres with a standard deviation of 70 kilometres. These compare favourably and are consistent with the values obtained earlier based on all the data.

The construction of high-rise buildings in Singapore accelerated from 1970. The responses of these high-rise buildings to the earthquake tremors from 1970 onwards are, therefore, most relevant to the situation in Singapore at present. Therefore, a detailed investigation on the reports of those major tremors is appropriate.

During the 1971, 1977, 1984 and 1986 tremors, many of the residents and office workers in the upper floors of high-rise buildings in certain locations were frightened and ran outdoors. There were a few incidences of cracked plaster but no structural damage was reported. Areas affected by these tremors are summarised in Table 11.2. It is apparent that only high-rise buildings in the south-eastern part of Singapore experienced earthquake tremors. There were no reports of buildings in other parts of Singapore experiencing similar tremors. Close examination of the location of those areas affected by tremors revealed that they were all located within the Kallang Formation area. This phenomenon will be explained in the following section.

Amplification of Seismic Waves

Earthquakes are vibrations of the earth's surface due to sudden movements of the earth's crust which consists of a number of solid rock plates floating on the molten rock interior of the earth. The plates, about 20 in total, drift on convection currents generated by hot spots deep within the earth. The movement is exceedingly slow and due to interlocking at plate boundaries, plates deforming at the boundaries, with the intra-plate areas remaining stable. As a result, stress builds up and when the shear stress exceeds the strength of the rock, a rupture occurs along the fault line or weak point in the rock and energy is released. This energy is dissipated in the form of propagation of waves in the material of the geological strata.

The seismic waves from Sumatra reaching the bedrock of Singapore comprise low amplitude long period acceleration components as high frequency components damp out within short distances from the epicentre. These waves are modified as they propagate through the overburdened soil of the bedrock underlying Singapore. As a result, the buildings are subjected to seismic waves of higher amplitude and longer period.

Seismic waves containing long period components will have a greater effect on buildings corresponding with natural periods in the neighbourhood of the dominant periods of seismic waves. Since long natural periods are associated with tall buildings, this class of buildings are excited by the seismic waves. In a vibrating building, the acceleration and displacement levels increase with the height. As a result, people in the upper floors notice the vibration more than those in the lower floors. Furthermore, the soils of the Kallang Formation are the most compressible among all the soil types prevalent in Singapore. Therefore, high-rise buildings located within the Kallang Formation will experience earthquake tremors more than the buildings in other locations.

In the design of high-rise buildings in Singapore, the lateral strength of most of the buildings is determined by the minimum value required by the building code, which is 1.5 per cent of the weight of the building. Because the prevailing wind condition is not so strong in Singapore, the lateral force acting on most of the buildings due to wind is usually less than the minimum value in the building code. Tremors, like the wind, induce lateral forces. The question is, whether this minimum lateral strength is adequate to resist the worst probable tremor.

Based on the earthquakes record from 1965 to the present as summarised in Table 11.1 and using the geophysical model developed by Balendra, Tan and Lee (1990), it is possible to check the adequacy of the code provision of lateral load of 1.5 per cent of the weight of the building for the design of high-rise building in Singapore.

The most common way of measuring the magnitude of an earthquake is the Richter scale. Effect of the ground motion due to an earthquake on the design of buildings is accounted for through the response spectrum. Generally, the recommendation for the design spectrum for the seismic analysis of nuclear structure given in Newmark, Blume and Kapur (1973) and the American Society of Civil Engineers Standards ASCE 4–86 (1986) is based on the probability level of (mean + standard deviation) of past record of earthquakes. No recommendation on the appropriate probability level with regard to the distance of the epicentre for the worst probable earthequake is given because they are concerned with near field effect of structures located within 100 kilometres from the epicentre only and not the far field effect as in the case of Singapore.

The local data on earthquake tremors available in Singapore do not permit the design spectrum to be established as the magnitude of the earthquakes in Sumatra was measured and reported in Richter scale. Nevertheless, for the magnitude of the worst probable earthquake intensity, it will be reasonable to adopt a similar probability level of (mean + standard deviation) of the past earthquake intensities. To be consistent in our design approach, it will also be reasonable to adopt a probability level of (mean-standard deviation) for the distance of the epicentre of the worst probable earthquake. It thus follows that, the worst probable earthquake tremor for Singapore will be due to an earthquake of 6.8 Richter occurring at about 500 kilometres from Singapore.

Table 11.3 summarises the relevant soil properties for each of the soil formations in Singapore compiled from Public Works Department (1976), Wallace, Ho and Broms (1990), Kurzeme, Hose and Nasim (1990), Poh, Chuah and Tan (1985) and Dames and Moore (1983). The thicknesses of each soil formation are chosen from the range reported to illustrate the effect of earthquake tremors on high-rise buildings. The corresponding results obtained from the geophysical model are also summarised in Table 11.3.

It is apparent from Table 11.3 that, in spite of amplification caused by the overburden soil strata, such an earthquake will produce a lateral force less than the empirical minimum value of 1.5 per cent specified in the code. Therefore, the lateral strength of the buildings in these parts of Singapore represented by the soil conditions illustrated in Table 11.3, designed according to the code of practice, is adequate to withstand earthquake tremors experienced in Singapore. However, this estimate is based on a geophysical model which has been validated using the attenuation of waves in the western part of the United States. As the attenuation characteristics are dependent on the type of rock in the path of propagation of waves, it is prudent to calibrate this analytical model using the attenuation data in this region, in order to use this model confidently. This will be possible only if seismograph stations are located in appropriate parts of Singapore. The information from the seismograph will prove to be useful in the future design and construction of high-rise buildings.

In addition, the present study reveals that the base shears in Jurong Formation and Old Alluvium as represented by the soil conditions in Table 11.3 are close to the minimum value of 1.5 per cent in the code of practice adopted in Singapore for the design of structures. It will be a good engineering practice to carry out similar analyses as demonstrated in the present study using a locally validated geophysical model when more critical soil conditions are encountered on site so that appropriate measures can be taken in the design of high-rise buildings.

Conclusion

Based on the limited statistical data available for Sumatra earthquakes and the geophysical model developed by Balendra, Tan and Lee (1990), it is found that the provision in the code of practice adopted for the design of structures in those parts of Singapore represented by soil conditions as in Table 11.3 is satisfactory. However, since the geophysical model has been validated with earthquakes in the western part of the United States, appropriate efforts should be made for the collection of data in this region to verify the model and enhance the limited statistical data available to date.

Human perception of discomfort during a tremor depends on the levels of acceleration and displacement experienced by the individual. When the threshold of discomfort is exceeded, movement is perceived by the individual which may cause the individual to panic. It has been found, in this present study that, this phenomenon is mostly associated with people living or working in high-rise buildings located within the Kallang Formation. This can be attributed to the high compressibility of the soils of the Kallang Formation, in comparison with the other formations, which enable a larger displacement to take place during a tremor.

Table 11.1 Summary of All Reported Earthquake Tremors in Singapore

Year	Date	Intensity of Tremor in Singapore (MM)	Richter Scale of Earthquake	Distance of Focus from Singapore (km)*	Near Field Intensity (MM) and Location	Source	Remarks
1986	August 12	VI	5.5	600	—	3, 4	
1985	December 27	II–III	6.5	—	—	3	
1984	November 17	V	7.4	500	—	6, 8	
1984	August 27	N.R.+	[4.1]	550	VII, Pahae, Jae and Tapanuli	1	Depth = 53 km
1983	August 31	IV	—	—	—	7	
1979	December 15	IV	[6–6.5]	550	VII–IX Bengkulu	1, 3, 5, 6	Depth = 25 km
1979	April 28	N.R.	[5.3]	450	V, Marapi	1	Depth = normal
1977	August 19	II	7.7	2200	VII, Sumbawa	1, 3	Depth = 33 km
1977	March 8	II	[5.5–6.5]	450	VIII, Pasaman	1, 2, 3, 5	Depth = 22 km
1976	July 28	I	—	—	—	2	The tremor may not be of earthquake origin
1976	June 20	N.R.	[6.0]	850	VII, Sibolga	1	Depth = 33 km
1975	January 8	II	[5.8]	550	—	2	Depth = 95 km
1972	December 18	I–II	[5.2]	400	—	2	
1971	February 4	IV	[7.2]	550	V–VI, Sibolga	1, 2, 3	*Straits Times* reported that the focus is 750 km from Singapore, depth = normal

Table 11.1 Summary of All Reported Earthquake Tremors in Singapore (cont.)

Year	Date	Intensity of Tremor in Singapore (MM)	Richter Scale of Earthquake	Distance of Focus from Singapore (km)*	Near Field Intensity (MM) and Location	Source	Remarks
1967	April 12	N.R.	[5.9]	850	VII, Aceh	1	Depth = 55 km
1965	July 25	N.R.	[4.8]	500	VII, Tapanuli	1	Depth = 98 km
1962	December 31	II–III	—	450	—	2	Depth = 67 km
1952	March 15	II	6.25 ROM	550	—	2	
1949	March 10	II	—	—	—	2	
1948	December 28	III	—	—	—	2	
1948	January 13	I–II	—	—	—	2	
1941	October 11	N.R.	[5.5]	700	VII, Tanapuli	1	
1938	August 18	N.R.	[5.5]	600	VII, Bengkulu	1	
1936	October 27	N.R.	[5.5]	600	VII, Tanapuli	1	
1936	September 9	III	[6.1] 7.2 PAS	750	VII, Tanapuli	1, 2	
1935	December 28	II	[5.5–6.1] 8.1 PAS	650	VII–VIII, Batu Island	1, 2	
1934	September 21	N.R.	[5.5]	500	VII, Tanapuli	1	
1933	June 25	N.R.	[6.1–6.7]	700	VIII–IX, South Sumatra	1	
1931	September 25	N.R.	[5.5–6.1]	—	VII–VIII, South Sumatra	1	

Table 11.1 Summary of All Reported Earthquake Tremors in Singapore (cont.)

Year	Date	Intensity of Tremor in Singapore (MM)	Richter Scale of Earthquake	Distance of Focus from Singapore (km)*	Near Field Intensity (MM) and Location	Source	Remarks
1926	June 28	III	[6.1–6.7] 6.50 PAS	500	VIII–IX, Singkarak	1, 2	
1922	February 7	II	—	—	—	2	
1922	January 31	II	—	—	—	2	
1921	April 1	N.R.	[6.7]	(660)	IX, Tanapuli	1	
1914	June 26	III–IV	[6.7] 7.6+	700	IX, Bengkulu	1, 2	
1909	June 3	IV	7.6+	500	—	2	
1904	June 15	N.R.	[6.1]	—	VII, Siri and Sori	1	
1902	June 27	N.R.	[4.9]	(600)	VI, Lais and Bengkulu	1	
1892	May 17	IV	[4.9]	(600)	VI, Prapat	1, 2	
1883	August 26–28	I	—	800	—	2	Eruption of Karakatoa
1873	October 7	N.R.	[4.9]	(600)	VI, Tanapuli	1	
1873	August 19	N.R.	[4.9]	—	VI, Mandailing	1	
1871	August 18	N.R.	[4.9]	(600)	VI–III, Bengkulu	1	

Table 11.1 Summary of All Reported Earthquake Tremors in Singapore (cont.)

Year	Date	Intensity of Tremor in Singapore (MM)	Richter Scale of Earthquake	Distance of Focus from Singapore (km)*	Near Field Intensity (MM) and Location	Source	Remarks
1861	February 16	IV	[6.1–6.7]	(600)	VIII–IX, Tapanuli and Sibolga	1, 2	
1852	January 9	N.R.	[5.5–6.1]	(500)	VII–VIII, Teluk Betung	1	
1843	January 6	I	[5.5–6.1]	(750)	VII–VIII, Nias Island	1, 2	Soetardji et al. reported earthquake occured on 5 January 1843
1837	—	—	—	—	—	2	
1835	August 26	N.R.	[5.5–6.1]	(450)	VII–VIII Padang	1	
1833	November 24–25	II–III	[6.1–6.7]	(600)	VIII–IX Bengkulu	1, 2	

N.R. indicates no reported tremor in Singapore.
* rounded off to the nearest 50 km.
[ ] indicates Richter scale obtained from body wave magnitude.
( ) indicates Richter scale and distance of focus from Singapore obtained from near field intensity of earthquake and location where near field intensity was reported.
\+ Scale used is not clear.

Sources

1. Soetardji et al. (1985)
2. Leyu et al. (1985)
3. *Straits Times*
3. *Straits Times*
4. *Business Times*
5. *New Nation*
6. *Sunday Times*
7. *Singapore Monitor*

Table 11.2 Areas Affected During Recent Major Tremors

Year of Tremor	Areas Affected
1971	Toa Payoh, Beach Road, Katong, Old Kallang Airport, Cecil Street
1977	Whampoa, Toa Payoh, Chai Chee, Marine Parade
1984	Beach Road, Rochor, Toa Payoh, Whampoa, Jalan Besar
1986	Robinson Road to Marine Parade, Beach Road, Jalan Besar, Toa Payoh

Table 11.3 Base Shear for Different Soil Conditions for an Earthquake of Magnitude 6.8 Richter at 500 km from Singapore

	Jurong Formation	Old Alluvium	Bukit Timah Granite	Kallang Formation
1. Undrained Modulus, E kN/m^2	200000	75000	10000	5000
2. Mass Density kg/m^2	2200	2000	1900	1500
3. Poisson's Ratio	0.2	0.2	0.3	0.45
4. Thickness (m)	90	100	25	15
5. Shear Modulus $G = E/2(1 + \upsilon)$ kN/m^2	83333	31250	3846	1678
6. Shear Wave Velocity (m/s)	195	125	50	33.5
7. Period (sec)	1.84	1.6	1.6	1.8
8. Hysteritic Damping	5%	5%	5%	5%
9. Base Shear as % of Weight	1.4	1.12	1.4	1.22

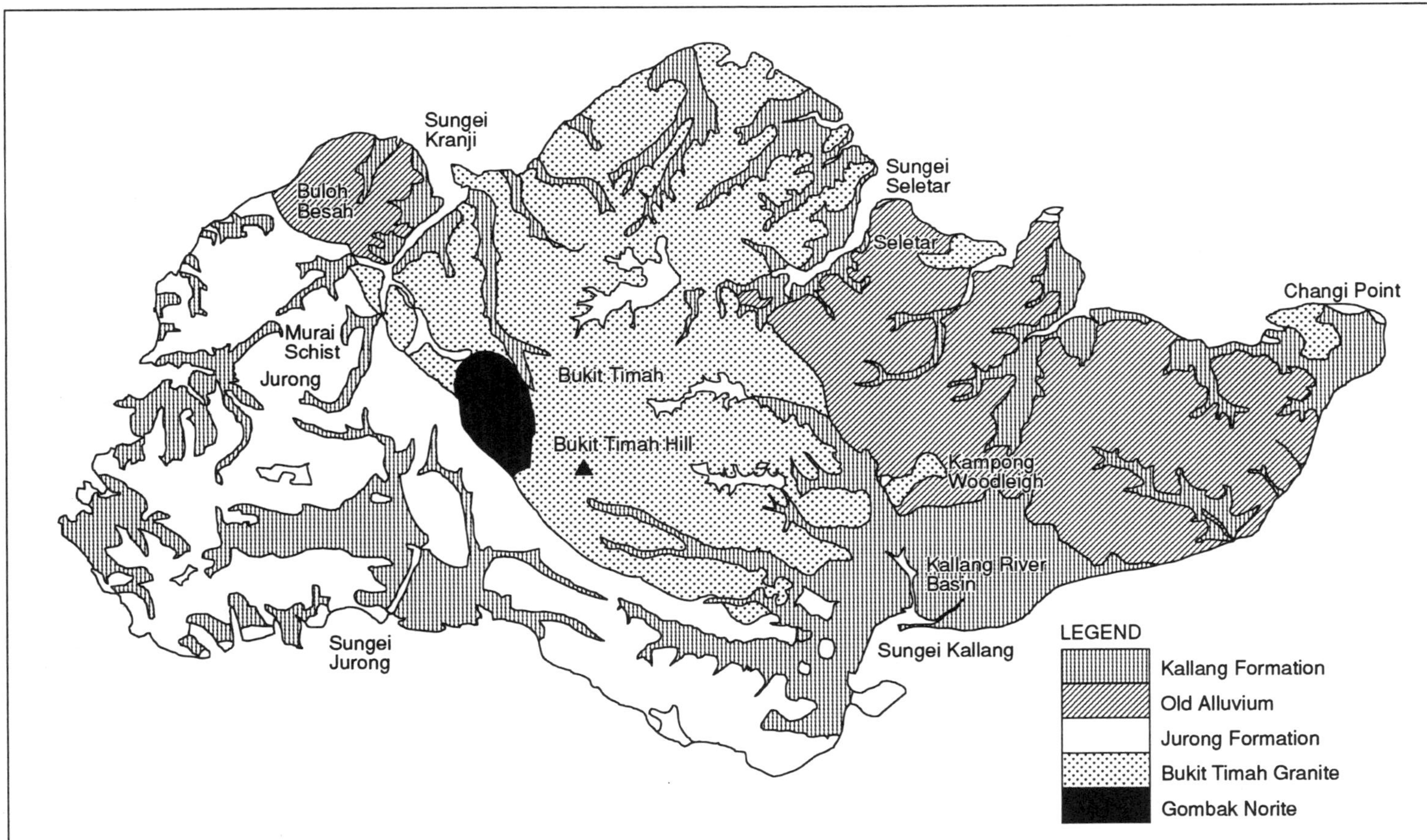

Source: Public Works Department (1976) *Geology of the Republic of Singapore, December,* Singapore.

Figure 11.1 Simplied Geological Map of Singapore Main Island

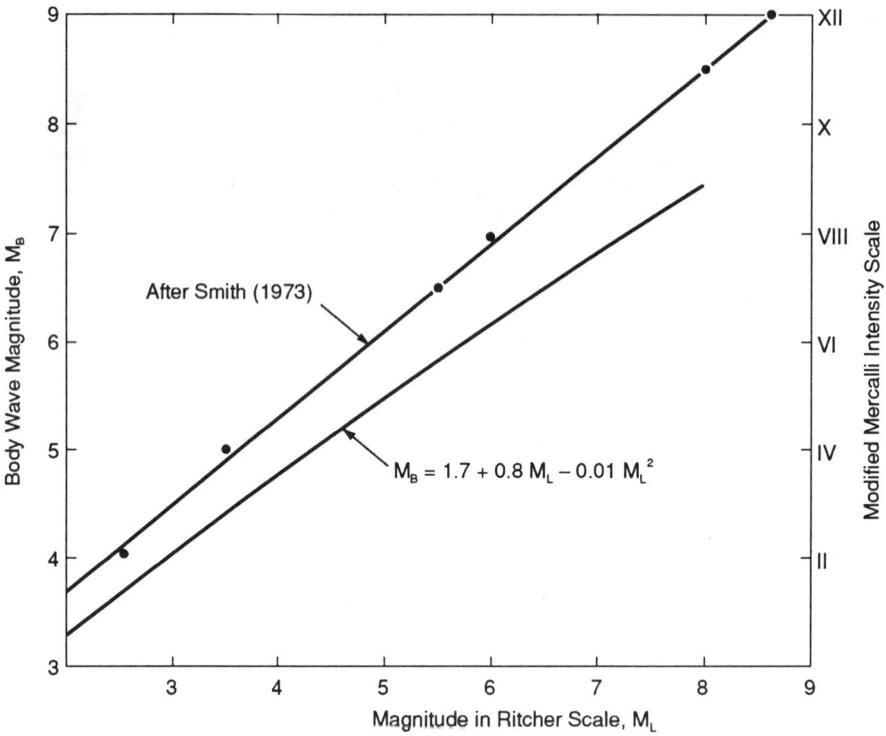

Figure 11.2 Relationship between Ritcher Scale, Body Wave Magnitude and Near field Modified Mercalli Intensity

APPENDIX

MODIFIED MERCALLI INTENSITY SCALE

I. Not felt.

II. Felt by persons at rest, on upper floors or favourably placed.

III. Felt indoors. Hanging objects swing. Vibration like passing of light trucks. Duration estimated. May not be recognised as an earthquake.

IV. Vibration like passing of heavy trucks or sensation of a jolt like a heavy ball striking the walls. Standing motor cars rock. Windows, dishes, doors rattle. Glasses clink. Wooden walls and frames creak.

V. Felt outdoors. Sleepers wakened. Liquids disturbed, some spilled. Small unstable objects displaced or upset. Doors swing, close, open. Shutters, pictures move. Pendulum clocks stop.

VI. Felt by all. Many frightened and run outdoors. Persons walk unsteadily. Windows, dishes, glassware broken. Knick-knacks, books, etc. off shelves. Pictures off walls, Furniture moved or overturned. Weak plaster and weak masonry cracked. Small bells ring (church, school). Trees, bushes shaken visibly.

VII. Difficult to stand. Noticed by drivers of motor cars. Furniture broken. Damage to weak masonry. Weak chimneys broken at roof line. Fall of plaster, loose bricks, stones, tiles, cornices, unbraced parapets and architectural ornaments. Waves on ponds. Small slides and caving in along sand of gravel banks. Large bells ring.

VIII. Steering of motor cars affected. Damage to ordinary quality masonry; partial collapse. Some damage to very good quality masonry; none to reinforced masonry. Fall of chimneys, monuments, towers, elevated tanks. Framed houses moved on foundations if not bolted down; loose panel walls thrown out. Branches broken from trees.

IX. General panic. Weak masonry destroyed; good quality masonry seriously damaged. General damage to foundations. Frames racked. Conspicuous cracks in ground.

X. Most masonry and frame structures destroyed with their foundations. Some well-built wooden structures and bridges destroyed. Serious damage to dams, dikes, embankments. Large landslides. Water thrown on banks of canals, rivers lakes etc. Rails bent slightly.

XI. Rails bent greatly. Underground pipelines completely out of service.

XII. Damage nearly total. Large rock masses displaced. Lines of sight and level distorted. Objects thrown in the air.

REFERENCES

ASCE 4–86 (1986) Seismic Analysis of Safety-related Nuclear Structures and Commentary on Standard for Seismic Analysis of Safety-related Nuclear Structures, American Society of Civil Engineers, Sep 86, 91 pp.

Balendra, T., Tan, T.S. and Lee, S.L. (1990) "An Analytical Model for Far Field Response Spectra with Soil Amplification Effects", *Engineering Structure*, Vol. 12, October, pp. 263–268.

Dames & Moore (1983) "Detailed Geotechnical Study — Interpretative Report, Mass Rapid Transit System", Singapore, p. 204.

Degg, M.R. (1992) "Natural Disasters — Recent Trends and Future Prospects", *Geography*, Vol. 77, No. 3, pp. 198–209.

Kurzeme M., Hose R.M. and Nasim S. (1990) "The OUB Centre Tower Foundation, Singapore", in *Proceedings of Conference on Deep Foundation Practice*, 30–31 October, Singapore, pp. 125–131.

Leung C.F. and Radhakrishnan R. (1990) "Geotechnical Properties of Weathered Sedimentary Rocks, Geotechnical Engineering", *Journal of Southeast Asian Geotechnical Society*, pp. 29–48.

Leyu C.H., Chang C.F., Arnold E.P., Kho S.L., Lim Y.T., Subramaniam M., Ong T. C., Tan C.K., Yap K.S., Shu Y.K. and Goh H.L. (1985) Series on Seismology, Vol. III — Malaysia, Southeast Asia Association of Seismology and Earthquake Engineering, June, 99 pp.

Newmark, N.M., Blume, J.A. and Kapur K.K. (1973) "Seismic Design Spectra for Nuclear Power Plants", *Journal of Power Division*, American Society of Civil Engineers, Vol. 99, pp. 287–303.

Poh K.B., Chuah H.L. and Tan S.B. (1985) "Residual Granite Soil of Singapore", in *Proceedings of Eighth Southeast Asian Geotechnical Conference*, 11–15 March, Kuala Lumpur, pp. 3–1 to 3–9.

Public Works Department (1976) *Geology of the Republic of Singapore*, December, Singapore, 79 pp.

Smith, Peter J. (1973) *Topics in Geophysics* (Cambridge, MA: Massachusetts Institute of Technology), pp. 166–176.

Soetardjo, M. Untung, Arnold E.P., Soetadi, R., Sulaeman Ismail and Engkon K. Kertapati (1985) Series on Seismology, Vol. V — Indonesia, Southeast Asia Association of Seismology and Earthquake Engineering, June, 199 pp.

Steedman, S. (1995) "Mega-cities — The Unacceptable Risk of Natural Disaster", *Built Environment*, Vol. 21, No. 2/3, pp. 89–93.

Wallace J.C., Ho C.E, and Broms B.B. (1990) "UOB Plaza Singapore Foundation Design and Construction", in *Proceedings of Conference on Deep Foundation Practice*, 30–31 October, Singapore, pp. 257–273.

Whittow, J.B. (1995) "Disaster Impact and the Built Environment", *Built Environment*, Vol. 21, No. 2/3, pp. 81–88.

12

Fear of Crime and Sense of Security in Public Housing Estates

Brenda S.A. Yeoh and Yeow Pei Lin

Introduction

Three and a half decades after its establishment, the Housing and Development Board's (HDB) efforts in "housing a nation" (Wong and Yeh 1985) has been held up by many as a success story. So successful are its programmes that the HDB won the Asian Management Award in Development Management in 1995 for "innovative sustainable and effective management" and substantially, improving the quality of life of the people (*Straits Times*, 17 October 1995). The Board's achievements are clearly inscribed in the built environment as each generation of New Towns and housing estates surpasses its predecessors in attention to aesthetic design and architectural detail, provision of a superior slate of social, recreational and retail amenities and signally, the higher market value of flats within the estates. This essentially progressive history of public housing in Singapore has been faithfully chronicled by HDB's own scribes as well as other commentators (see, for example, *Homes for the People* 1965/66; Yeh 1975; Hassan 1977; Teo and Savage 1985; Wong and Yeh 1985).

More recently, analysts have more explicitly drawn out the ideological intents of the public housing programme in Singapore, making clear that the programme not only addresses the functional problem of providing Singaporeans a roof over their heads but also facilitates the achievement of state-vaunted ideological goals such as promoting economic development, enhancing social integration, perpetuating desired social and moral values and legitimising the ruling party's political domination (see, for example, Castells, Goh and Kwok 1990; Chua 1991; Ooi 1991, 1995; Hill and Lian 1995).

While many of the above works have moved beyond examining the brick and mortar of public housing to interpreting the moulding of community life and associated management of political, economic, cultural and ethnic dimensions in the context of a specific built environment, much less has been said about the dysfunctional aspects of living in a public housing estate. A few commentators have pointed out that there are occasional slippage and contradictions between the HDB's vision and the actual effects which have disadvantaged certain less privileged groups such as the poor and the elderly (Salaff 1988; Yeoh and Kong 1994; Blake, this volume).

Another aspect of HDB living, one which has attracted little comment but is crucial to the quotidian well-being of HDB inhabitants, is the question of the extent to which, crime, violence and fear are prevailing concerns in an HDB environment. In one sense, Singapore is well-known today as a safe city where public streets and places are free of violence and crime. Residential scenes of the 1950s and 1960s typified by urban slum areas ridden with various forms of social ills associated with secret society activities such as gang fights, armed robberies, intimidation and extortion, blackmail and the collection of protection money, kidnapping, hooliganism, rape, prostitution, murder, gambling dens and juvenile delinquency (Blythe 1969; Mak 1981) seem to have been excised from the landscape as a result of comprehensive urban renewal and the concomitant rehousing of people in New Towns. The HDB environment is not only more physically ordered and sanitised, but also serves as a means of purifying the once socially-polluted landscape.

Crime statistics setting Singapore against other western and eastern cities today support Singapore's crime-free image.[1] Yet, in Singapore's highrise residential environments, crime is not entirely extinct. A 1984 report, for example, shows that, while crimes committed in public housing estates "do not give cause for alarm", certain types of crimes such as lift robberies are on the increase and that most of these incidence actually occur during the day time, a phenomenon which was attributed to the fact that "most of the men folks [*sic*] are at work and the housing estates are relatively quiet" (*Report and Recommendations of the Public Housing Sub-Committee* 1984, p. 5).

More crucially, it must be noted that, the spatial experience of *fear* need not coincide with the actual map of *crime* and while in Singapore, there is no "climate of fear" of "alarmingly pervasive" proportions as in the case of the United States (Quah and Quah 1987, p. 33), surveys from a number of countries have consistently shown that women expressed distinctly higher levels of fear of crime than men (Clemente and Kleiman 1977; Gordon et al. 1980; Riger and Gordon 1981; Lagrange and Ferraro 1989; Valentine 1989; 1992; Yeow 1993/94; Chang 1994; Currie 1994). Although women may suffer equal, if not lower risk of victimisation compared to men, their greater fear of crime has been attributed to one crime — rape — the single offence which primarily victimises women and which heightens their fear of victimisation by other crimes (Warr 1985). As such, in the public realm, women's fear of crime becomes a fear of men, or more specifically, male strangers. Since public space forms the context where strangers meet and where men dominate, women's fear of male strangers is transposed to unease within the public arena (Valentine 1989).

In this paper, we focus on evaluating the question of personal security within public housing estates in Singapore with specific emphasis on the perceptions and experiences of women. The aim is to discuss how a stronger sense of security that takes into consideration women's sensitivities to environmental cues of danger can be fostered within public housing estates. We will begin with a brief discussion of first, the links between crime and the environment; and second, existing crime prevention measures in public housing estates.

Approaches to Crime and the Environment

The causation and treatment of crime, a multi-faceted phenomenon, cannot be pinned down by any single-stranded theory. Hence, attempts to understand crime have taken on a multi-disciplinary approach (Herbert 1977, p. 208). Criminological works that focus on the relationship between crime and the environment can be broadly divided into two perspectives — "offender"- and "victim"- (or "event") centred approaches.

Among offender-centred approaches, sociological theories which relate the offender to the immediate social environment include *inter alia*, social disorganisation theories (Sutherland 1939; Durkheim 1952; Shaw and McKay 1969) and subcultural studies (Thrasher 1928; Matza 1964; Jankowski 1991). In general, the former suggests that it is the absence of a stable form of society with clearly defined behaviour codes and established norms which precipitates conditions for deviance. Poverty, poor housing and unemployment amidst rapid social and economic change indicate social disorganisation and are linked to the prevalence of crime. Under the rubric of subcultural studies is a loose agglomeration of theories which essentially conceive of delinquency and deviance as products of systematic learning of values and norms which are discordant to those of the mainstream society. These theories employ concepts of "turfs" and "territories" which can be extended to other more loosely defined collectivities such as the neighbourhood and other environmental contexts (Herbert 1977, p. 211).

Geographical perspectives on crime have even more strongly emphasised the spatial and environmental contexts in which crime occurs. By mapping locations where crime occurs and the movement patterns of criminals when picking up targets, crime is established as a place-specific activity that sometimes shows marked clustering in particular parts of the city (Brantingham and Brantingham 1984, p. 212; Herbert 1977, p. 209). Increasingly, a more radical perspective has found expression in the study of the geography of crime where the emphasis has shifted from a concern with the local environment to an analysis of capitalist structures and social relations (Peet 1975; 1976; Smith 1983; 1987; 1989).

When researchers realise the importance of situational factors in crime occurrences, they began to study crime from a victim-centred perspective. Research in this field — victimology as it is often called — focuses on the re-enactment of the criminal event, the victim and the location of the offence. Key strands of this work include *inter alia*, studies on fear of crime and those examining crime prevention through urban design and planning.

Studies on fear of crime are mainly concerned with identifying who the "fearful" individuals are by correlating expressed fear levels with various socio-demographic variables such as age, gender, ethnicity, income, education level and place of residence (Biderman et al. 1967; Goldsmith and Goldsmith 1976; Clemente and Kleiman 1977; Gordon et al. 1980; Yin 1980; 1982; Lagrange and Ferraro 1989). An extension of such work is to examine the social and spatial consequences of fear on individuals.

One example, as has been generally noted, would be that those who feel unsafe are more likely to avoid going out on foot (Smith 1987, p. 12) and that these restrictions are especially acute for women and the elderly. In another example, Lawton and Yaffe (1980) noted that fear of crime leads the elderly to restrict the frequency with which they go out and this, in turn, isolate them socially and affect their psychological well-being while Valentine (1989) argues that women's use of precautionary measures and avoidance strategies restrict their lifestyles, access to public space and participation in public life.

The environmental context is given specific emphasis in works concerned with the relation between crime and urban design. Pioneering efforts in this area are credited to Jacobs (1962) and Newman (1972). Crime is, basically, an act that needs the presence of an offender, a victim and a place. While it is important to understand why an offender commits a crime, it is also necessary to examine the context of the crime in which the three elements come together. The built environment, by virtue of its design, can directly or indirectly — through its influence on the social environment, create opportunities for would-be offenders to commit crimes.

In Jacobs' (1962) *The Death and Life of Great American Cities*, she drew attention to the relationship between crime prevention and the design of the built environment. She argues that the public peace of city streets is not kept, primarily, by the police but by an intricate network of informal social controls made up of the presence of the people themselves — mainly strangers who act as "eyes upon the street". Specific measures such as allowing a good mix of landuses to create as many busy areas as possible can thus help to enhance public surveillance (Jacobs 1962, pp. 35–37).

Similarly, Newman (1972) centred his thesis on the importance of urban design in creating differential opportunities for informal surveillance. His key contribution lies in theorising about "defensible space", a space that physically expresses a social fabric which defends itself, defined by the concepts of **territoriality**, **natural surveillance**, **image** and **milieu**. A residential estate with good defensible space must be able to create clear zones of perceived **territorial influence**. Such an environment that encourages inhabitants to assume rights of control over spaces will be more secure. The design must also provide the residents with **natural surveillance** opportunities of the surroundings. Where non-private spaces can be casually and continually surveyed, security will be improved because intruders and victims can be readily identified and help rendered when necessary. By **image**, Newman refers to the use of physical design to avoid stigma. A well-designed place will promote the residents' sense of attachment to their environment and avoid stigmatisation and consequent alienation that accompanies a dilapidated appearance of the project. Finally, the attribute **milieu** is concerned with juxtaposing projects with "safe zones" in adjacent areas. More specific studies have also identified "surveillability" (i.e. cover, presence and proximity of neighbours) as the most important factors influencing offender's choice of targets (see Stollard 1991, pp. 7–8 for a review).

In an attempt to integrate earlier findings, Rand (1979) outlines four theories of crime prevention through environmental design: social control (i.e. unconscious surveillance); enclose/access control (i.e. the use of definite boundaries to discourage or prevent criminal access); criminal justice (i.e. the presence of a security force); and defensible space (borrowing Newman's idea).

While these theories have been criticised for their environmentally deterministic stance and their neglect of social factors such as community ties in influencing the level of informal control (Mawby 1974; Mayhew 1981), they have identified an important theme. The theme, more specifically, is the importance of features of the environment which facilitate informal social control, given that, in public places, people usually feel safer in the visual presence of many others and in general, the relationship between crime, fear and the design of the built environment.

Crime Prevention Measures in Public Housing Estates

In the past, it was generally felt that the responsibility for crime prevention lay with the police and the individual householder. In Singapore, the role of the police is essentially to maintain law and order, to preserve the public peace, to prevent and detect crime and to apprehend offenders (Quah and Quah 1987, p. 2). However, it has been observed that deploying more police officers alone does not necessarily lead to lower crime rates and that, the police often do not have the resources to intervene in various circumstances which result in crime being committed (Stollard 1991, p. 9).

In the 1980s, community-oriented policing was adopted as a basic strategy to fight crime in Singapore, particularly in public housing estates. According to Quah and Quah (1987, p. 5), one of the reasons was the decentralisation of the population and rehousing in high-rise flats in New Towns. Consequently, the former method of "vertical policing" (i.e. policing each level) was not only no longer feasible but also ineffective in providing sufficient opportunities for interaction between residents and policemen on patrol. Changing housing design thus necessitated new strategies of policing.

The beginnings of community-oriented policing was signalled by the setting up of the National Crime Prevention Council in 1981 to promote public awareness of crime and "to mobilise a wider section of the public to actively partcipate in the work of crime prevention" (Chua 1984). This was followed by the initiation of various projects such as the Neighbourhood Watch Scheme, crime prevention campaigns and boys' clubs and other organised activities to channel the young towards healthy pursuits and away from crime (Tng 1992). Its main thrust is, however, encapsulated in the Neighbourhood Police Post (NPP) system, an adaptation of the *koban* system in Japan. Based on the twin aims of improving police-community relations and preventing and suppressing crime through the cooperation and support of the community, a network of NPPs or mini-police stations were set up in central locations in housing estates, each catering to about 30 000 residents (Bayley 1983; Quah and Quah 1987, p. 7).

The importance of improving police-public cooperation and the success of the NPP system in Singapore is well-attested to (Ong, Mak and Chang 1982; Tan 1988; Quah and Quah 1987; Loke 1994). While involving the community is a crucial step, a broader approach to crime prevention would also include attention to urban design features as part of a larger package of safety improvements. The complexity of crime as a phenomenon warrants a multi-pronged approach combining the roles and resources of the police, Town Councils, housing authorities and neighbourhood organisations, the community of people as well as a defensibly designed environment which creates "multiple opportunities for [informal] surveillance" and encourages "patterns of mutual accountability" (Stollard 1991, p. 17).

In what that follows, we examine the physical characteristics of public spaces in one of the premier public housing estates in Singapore — Bishan New Town — in relation to the opportunities they create for personal (as opposed to property) crimes, in particular, rape and molestation, crimes which affect mainly women. This provides the local context to discuss principles for working towards a defensibly designed environment in public housing estates.

Environmental Peril in a Public Housing Estate

Women[2] identify places of environmental peril in Bishan New Town in two broad categories. The first category would be open spaces such as Bishan Park and its adjacent areas (canals, streets, jogging and cycling tracks) and construction sites; and second, enclosed and semi-enclosed spaces including multi-storey car parks, void decks, corridors, staircases and lifts. Their geography of fear also has a temporal element in that these spaces are seen to be more dangerous at night.

Bishan Park, one of the largest parks located in an HDB estate, boasts a wide range of recreational facilities such as a fishing pond, bicycle kiosks, jogging and cycling tracks, children's playgrounds and an amphitheatre and stage. It is potentially the most delightful part of the estate in both aesthetic and recreational terms; yet, in many women's eyes, it tops the list of dangerous places in women's mental maps of fear.

The most frequently cited reason for women's unease is the poor utilisation of the park at night. Peak hours, when the park is in use are between 5 pm and 7 pm. Before 5 pm, the place is simply a deserted, dispirited vacuum save for a few lone joggers, and mothers with children and elderly folk taking a stroll. Once dark, the park is again, largely unoccupied as the number of people dwindles to a few male joggers and courting couples. No female park user, whether alone or in groups, can be spotted. Once devoid of people, the park becomes a vast, dangerous terrain removed from the physical presence and potential bystander intervention of others. Such a landscape of fear is further compounded by newspaper accounts and hearsay which portray the park after dusk, as a paradise for perverts and muggers.

Women also point out that certain design features in the park impair clear visibility and hence surveillance of on-going activities. Dense landscaping such as bushes above one metre in height and clusters of shrubs found at junctions of footpaths and other pausing places such as shelters intended to give the park a sense of aesthetic intricacy are more perilous than pleasant at night. In the words of one respondent: "there are dense clusters of trees and bushes all over ... it makes me feel very worried when I walk past them because someone could easily hide in there". Other unpopular features that constrain visual access include meandering paths and the sloping gradients of the ground. The inability to see what is round the corner or on the other side of a pathway creates a sense of apprehension amongst women.

The problem of visibility posed by landscaping is aggravated by inadequate lighting. Parts of the park which are commonly singled out by women as dangerous grounds, included the multi-purpose turfed area and the jogging track running parallel to a canal leading to the park proper as there is a complete lack of lighting in these areas. Women feel even more vulnerable when the lack of "visibility by presence" is further diminished by deficiencies in illumination (Maitland 1992).

Further, the women also share the common sentiment that the scale of the park is too large for effective natural surveillance.

While the design of the park incorporates one of Newman's (1972) defensible space guidelines of juxtaposing an unsafe activity area (the park) and safe zones such as busy roads with flats overlooking the park, the sheer size of the park itself inhibits surveillance. At night, given the dim lights of the park, casual surveillance from surrounding flats or roads is made all the more impossible. The "unhuman" scale of the park relative to the number of users accentuates its emptiness and hence, perceived danger. According to one respondent:

> there is only one part of the park that is well-used ... the part where the big pond for fishing is ... the park is just too big for my liking ... there are so few park users, especially at night ... the place is so empty and lifeless.

In addition, the surrounding land uses, in particular, the canals and adjacent streets also become associated with the unsafe environment of the park in women's minds. This is noted by Jacobs (1962):

> ... unpopular parks are troubling because they have the same problems as streets without eyes and their dangers spill over into areas surrounding, so that the streets along such parks become known as danger places too and are avoided.

Sites adjacent to the park are thus included in the list of places women would rather avoid. In the words of one respondent:

> I try to walk on the other side of the road because this stretch of road next to the park is so quiet.

At the time of the survey, two construction sites were singled out by Bishan women as being unsafe places. These are located next to the Bishan Mass Rapid Transit station and the bus terminus respectively. Again, poor lighting within and outside these sites is seen as a danger cue. Furthermore, the undeveloped and isolated nature of these areas connotes that help would be unlikely to come by if one were attacked. For the site located next to the bus terminus, the dense growth of trees and bushes also creates hiding places for potential criminals. "Signs of incivility" (Valentine 1992) such as litter and disorder in what is seen as "wasteground" as well as the perception that construction sites are controlled by an alien or a threatening group also raises the fear index for construction sites. As one respondent who lives opposite a construction site puts it:

> I would not let my daughter go near there ... those [construction] workers look so lecherous and aggressive ... They (the contractors) did not even bother to fence up the place ...

Hence, aside from the problem of visibility, these sites are regularly stigmatised as a hotbed of criminal activity, a belief that stems from a general distrust of foreign construction workers. Women's perceptions of safe and unsafe areas are thus, dependent not only on the cues they receive from the physical environment but also the preconceived images they have of particular areas and occupants.

Bishan is one of the first few public housing estates to have multi-storey car parks alongside surface ones. From the HDB's perspective, multi-storey car parks make for intensive development of scarce land and optimum land use, leaving, for instance, enough space for landscaped areas and playgrounds to be built (*Straits Times*, 25 January 1994). For most women interviewed, however, they do not meet safety needs and are shunned unless one is unable to find a parking lot in nearby surface car parks. Multi-storey car parks are seen as hazardous for three reasons. First, there is the absence of human traffic save for motorists. One respondent feels that:

> It is so quiet in there ... Once, when I came back late, there was no one around but me and I was really, really scared, imagining the craziest of things ... Since that time, I do not use the [multi-storey] car park after 9 pm. I rather park my car further away in a surface car park a few blocks away.

Surface car parks are, indeed, safer as far as the security of being able "to see and be seen" is concerned. Unlike open-air lots where there is a constant stream of pedestrians on their way to bus-stops, shops or homes, the multi-storey car park is isolated from such traffic on the ground. In public housing estates, these car parks operate primarily on the basis of season parking. Thus, unlike those in public or commercial buildings, where there is at least one car park attendant around, no such permanent staff are stationed at the entrances or exits who could render help in case of need.

Second, respondents also feel insecure because the existing design features of multi-storey car parks place users out of the visual range of others. One respondent says:

> The place is totally unsafe, an absolute nightmare to be in ... it is so dark, so enclosed ... [As for] those pillars, I keep thinking that someone is hiding behind them ...

More specifically, the women single out four obstacles to long-distance view. Passers-by are unable to look into the car parks and note any extraordinary happenings, especially on the higher floors. The numerous pillars and cars also present opportunities for lone criminals to conceal themselves and strike unobserved. Furthermore, these car parks are generally perceived to be insufficiently lit. Finally, the women are uncomfortable with the fact that the staircases which lead out of these car parks are hidden from public view. In short, poor visual access and the absence of official staff and passers-by imply the dangerous absence of public surveillance and policing.

Third, the enclosed nature of multi-storey car parks with their limited entrances and exits creates difficulties for users to escape when attacked. This is graphically described by one of the respondents:

> From the time I park my car till the minute I leave, every moment is just filled with so much tension ... there are so many corners, pillars ... [and] with place so enclosed ... it will be difficult to try to run out quickly.

To many respondents, the areas just outside of the home — corridors, staircases, lifts and void decks are spaces fraught with possible danger. Reasons for such perceived danger can be found in Jacobs' (1962) analogy of these semi-public spaces as "figurative streets". In her view, these spaces are streets in the sense that first, they serve the comings and goings of residents, most of whom (particularly in a new town of rather shallow depth in terms of length of residence and the degree of community interaction among inhabitants) are strangers to one another; and second, they are accessible to the public. Anyone can come and go as they please as public housing lack symbols of territorial control such as gates, identification cards, security checks and electronic interview systems integral to private apartment blocks. In other words, while these spaces are semi-private as they are meant primarily for residents' use, their design in terms of physical accessibility without an attendant official means of spatial control confers on them a somewhat public character. In the words of one respondent:

> Anyone can just walk right up to the flat. There is no way I can tell whether he stays around here or not ... Not to mention that I do not know many of my neighbours ... Even though I am near home, I do not feel safe until I step into the house.

In terms of Newman's (1972) concept of territoriality, these spaces are, essentially, no-one's turf and therefore, all responsibility for the activities outside the immediate confines of individual apartments are relegated to public authorities. Theoretically, informal control can still develop. While strangers, on the one hand, threaten residents' ability to control access to semi-public spaces, they can also, on the other hand, add to the number of "eyes" in the "streets". In other words, the free access of space can be turned into a security asset on the principle that well-meaning strangers police predatory ones. More often than not, however, this potential asset turns out more to be a liability because of poor visual access and the low and unpredictable volume of pedestrian traffic. Staircases, corridors and lifts are more akin to enclosed or semi-enclosed spaces hidden from public view from the ground floor. One respondent says:

> In the daytime. there is hardly anyone around. My neighbours have either gone to work or to school. If anything should happen ... even just outside my door, passers-by cannot possibly see.

Even within each individual floor, there are places for concealment such as corners of staircases, lift lobbies and the pillars of void decks. This is further compounded by inadequate lighting along some corridors and in certain void decks. Some women also feel that the lifts are poorly sited in terms of visibility. One respondent comments that:

> The lift in this block faces a blank wall instead of the road. This is not very safe since someone may just grab you into the lift and nobody will know.

Evidently, streets in the sky lack the checks and inhibitions exerted by eye-policed city streets. They are, in Jacobs's (1962) terms, "blind-eyed streets". Like other enclosed and semi-enclosed spaces, the design of corridors, staircases, void decks and lifts easily allow an offender to trap his/her victim.

Towards a Safer Living Environment: Principles and Practices

It is clear from the above discussion that, women generally feel more secure in places that are designed such that informal control is perceived to be in effect by virtue of the actual or potential presence of others and where entrapment by the opportunist offender is made difficult. One of the most important principles in reducing fear and improving safety concerns natural surveillance of both dwellings and surrounding public spaces. The idea of surveillance can be applied not just to individual homes but also public places such as parks and playgrounds, parking lots, footpaths and access points to blocks of flats. For example, lift landings or lobbies should be designed in such a way that they are protected by natural surveillance. They should be located such that they do not face blank walls or are concealed in some way. Corners and bends should be avoided. An example would be the design of the entrance to a lift such that one does not need to make a turn and enter the lobby "blindly", not knowing what is ahead.

As a place where one often comes in close proximity with strangers within an enclosed space, the lift is fraught with dangers. Recent efforts to install viewing windows so as to allow passers-by to look into the lift interior is a step in the right direction. This is encouraging natural surveillance[3] although this measure is currently restricted to selected blocks of flats.

Other communal space within blocks of flats such as the letter-box area should preferably open up to streets or pathways so that internal activities such as retrieving mail from letter-boxes are observable by passers-by. It has also been suggested that the circular, U- or L-shaped block has a lower crime rate than the slab block with long common corridors "because the windows of flats [in the former] face inwards thereby creating a psychological fear in the would-be criminal that he [*sic*] is being watched" (*Report and Recommendations of the Public Housing Sub-Committee* 1984, p. 18).

Covered multi-storey car parks with unassigned lots are often perceived to be perilous as they go against the logic of casual surveillance. Women often prefer on-street parking close to dwellings or adjacent to busy areas. However, granted that multi-storey facilities are essential under certain circumstances such as where there is a need to economise on the use of land, the management of these car parks could be improved to provide women with a greater sense of security. Measures could be taken by the HDB using the revenues it collects from car parking charges such as employing permanent car park attendants at night or using electronic security systems to cover parking areas.

In public parks and playgrounds, landscaping should not unnecessarily detract from users' visibility, nor create secluded areas where intruders can hide. Apart from landscaping considerations, a sense of security is often enhanced if the playground is overlooked by a number of dwellings or if the park attracts the presence of a sustained flow of users.

In the case of Bishan Park, its size and lack of activities renders it under-used, particularly, in the evenings, and where public space is devoid of people, it simultaneously becomes an unsafe place as well. Its unpopularity outside of the "5 pm to 7 pm" peak-hour is largely a functional problem and is specifically tied to the lack of specialised leisure activities and attractions. Most women agree that, on the whole, the facilities provided by smaller neighbourhood parks such as jogging tracks, play-grounds and exercise areas are sufficient to meet the leisure needs of women and others.

Bishan Park, despite its advantages in terms of size and aesthetic qualities, does not provide any specialised activities that would differentiate it from other neighbourhood parks. A strategy aimed at creating a distinctive identity for the park and also to attract more users would be to redevelop the park for a calendar of distinctive activities such as regular concerts, cultural events and carnivals. "After-five" eating places can also be set up in the park to provide an alternative for out-of-town diners and also to draw crowds. Enlivened in such a fashion, the park would also benefit from a diverse rim of activities in the neighbouring hinterland.

As noted by Jacobs (1962), the park "is the creature of its surrounds and of the way its surroundings generate mutual support from diverse uses, or fail to generate such support". Hence, effort should be made to nurture a neighbourhood in Bishan where life swirls, where commercial, work and cultural activities enrich residential land use. In doing so, a wide array of users will be drawn to populate and animate the park throughout the day and during the evenings.

Conclusion

Design features are likely to be effective only when they complement other means of crime prevention, particularly, those pertaining to good housing management as well as those which enhance social cohesion by creating a "community of interests" (Poyner 1983) where people identify with one another. Despite the term, "natural" surveillance is not entirely spontaneous but at least partially socialised, and design and other social measures which work towards encouraging a sense of community and stronger social interaction among neighbours, will contribute towards making intruders feel conspicuous, since the residents will be more able to identify "outsiders" and more willing to look out for one another. In this regard, a sense of security and effective surveillance are also predicated on the residents' sense of place and territoriality.

While the HDB precincts in many New Towns are designed to enhance this idea of territoriality by creating semi-enclosed spaces with a focal point for activity (usually a playground or hardcourt meant as gathering places for residents), the resultant sense of community varies from estate to estate. In "vintage" estates such as in the former Singapore Improvement Trust estate in Tiong Bahru, social cohesion drawing sustenance from historical roots, formal and informal networks, the distinctiveness of the built environment and a lattice of everyday routines is strong (Yeoh and Kong 1994).

Similarly in a mature HDB New Town like Ang Mo Kio, Ooi (1995) reports that specific neighbourhood sites and facilities "provide opportunities for frequent routinised exchanges [which] have contributed to greater familiarity with other residents over time and a stronger sense of community". In contrast, neighbourly ties are much more poorly developed in new housing estates such as Pasir Ris (*Report and Recommendations of the Public Housing Sub-Committee* 1984, p. 7; Teo and Huang 1996), making the defence of space against intruders less effective. Where successful, the state's concerted efforts in building a sense of community through various grassroots organisations such as Residents' Committees, Civil Defence Coordinating Committees and Community Clubs thus, have important spin-offs in nurturing a sense of security.

In combating the fear of crime, environmental design alone is also inadequate in another sense where women are concerned. Insofar as women are socialised to perceive themselves as physically weak and to regard the confines of the home as women's rightful place while men dominate public spaces and wield the threat of rape as a form of social control, women will continue to feel vulnerable to sexual assaults (MATRIX 1984; McDowell 1989; Little 1994). The environment is merely the context onto which women transpose their fear.

Ultimately, for fear to be effectively alleviated, fundamental questions relating to a society's notions of masculinity and femininity and the gender inequalities which structure society must also be addressed. This, however, is a much more long-term resolution whose salience does not necessarily militate against other fear alleviating and crime prevention measures, including working towards a more defensibly designed built environment in our public housing estates.

ENDNOTES

1 See *Straits Times*, 23 November 1994, which reported that in 1991, Singapore's rate of violent crimes such as murder, rape,

robbery and aggravated assault was the lowest among a number of cities with the exception of Tokyo. It also reported on a survey which showed that most Singaporeans view the island as a "safe and secure" place characterised by law and order. See also Yip and Ong (1991).

2 The discussion is based on a survey of 150 women and in-depth interviews with 15 selected respondents on their experiences of fear and crime in Bishan housing estate.

3 An earlier experiment with using the CCTV lift security monitoring system to achieve a level of survelliance was discontinued after the pilot phase (*Report and Recommendations of the Public Housing Sub-Committee* 1984).

REFERENCES

Bayley, D.H. (1983) "The Significance of Singapore's Experiment with Neighbourhood Police Posts", *Police Life Annual*, pp. 44–49.

Biderman, Albert D., Louise Johnson, Jenny McIntyre and Adrianne Weir (1967) *Report on a Pilot Study in the District of Columbia on Victimization and Attitude toward Law Enforcement and Administration of Justice* (Washington D.C.: U.S. Government Printing Office).

Blythe, W. (1969) *The Impact of Chinese Secret Societies in Malaya — A Historical Study* (London: Oxford University Press).

Brantingham P.J. and Brantingam P.L. (1984) *Patterns of Crime* (New York: Macmillan Press).

Castells, M., Goh, L. and Kwok, R.Y.W. (1990) *The Shek Kip Mei Syndrome — Economic Development and Public Housing in Hong Kong and Singapore* (London: Pion).

Chang, H.Y. (1994) "Crime in Singapore — A Statistical Comparison with Major Cities", *Statistical Singapore Newsletter*, 17(2): 2–6.

Chua B.H. (1991) "Not Depoliticized but Ideologically Successful: the Public Housing Programme in Singapore", *International Journal of Urban and Regional Research*, 15: 24–41.

Chua S.C. (1984) "Speech by Mr Chua Sian Chin, Minister for Home Affairs, at the Opening of the NCPC Seminar "Crime Prevention — Everyone's Interest and Responsibility", RELC International House, 14 July, Singapore.

Clemente, F. and Kleiman, M. (1977) "Fear of Crime in the United States — A Multivariate Analysis", *Social Forces*, 56: 519–31.

Currie, D.H. (1994) "Women's Safety on Campus — Challenging the University as Gendered Space", *Humanity & Society*, 18(3): 24–48.

Durkheim, E. (1952) *Suicide — A Study in Sociology* (London: Routledge and Kegan Paul).

Goldsmith, J. and Goldsmith, S. (1976) "Crime and the Elderly — An Overview", in Goldsmith, J. and Goldsmith, S. (eds.) *Crime and the Elderly* (Lexington: Lexington Books).

Gordon, M.T., Riger, S., LeBailley, R.K. and Heath, L. (1980) "Crime, Women and the Quality of Urban Life", *Signs*, 5: 144–60.

Hassan, R. (1977) *Families in Flats — A Study in Low Income Families in Public Housing* (Singapore: Singapore University Press).

Herbert, D. (1977) "Crime, Delinquency and the Urban Environment", *Progress in Human Geography*, 1: 208–39.

Hill, M. and Lian K.F. (1995) *The Politics of Nation Building and Citizenship in Singapore* (London: Routledge).

Homes for the People — A Review of Public Housing (1965/66) (Singapore: Straits Times Press).

Jacobs, J. (1962) *The Death and Life of Great American Cities* (New York: Random House).

Jankowski, M.S. (1991) *Islands in the Street — Gangs and American Urban Society* (Berkeley: University of California Press).

Lagrange, R. and Ferraro, K. (1989) "Assessing Age and Gender Differences in Perceived Risk and Fear of Crime", *Criminology*, 27: 697–719.

Lawton, M. and Yaffe, S. (1980) "Victimisation and Fear of Crime in Elderly Public Housing Tenants", *Journal of Gerontology*, 35: 768–79.

Little, J. (1994) *Gender, Planning and the Policy Process* (Oxford: Pergamon).

Loke, W.H. (1994) "Living in One of the Safest Cities in the World", *Police Life Annual*, pp. 76–82.

Maitland, D. (1992) "Women, Space, Vulnerability and Safety", *The Operational Geographer*, 9: 28–29.

Mak, L.K. (1981) *The Sociology of Secret Societies — A Study of Chinese Secret Societies in Singapore and Peninsular Malaysia* (Kuala Lumpur: Oxford University Press).

MATRIX (1984) *Making Space — Women and the Man-made Environment* (London: Pluto).

Matza, D. (1964) *Delinquency and Drift* (New York: John Wiley).

Mawby, R. (1974) "Defensible Space — A Theoretical and Empirical Approach", *Urban Studies*, 14: 169–79.

Mayhew, P. (1981) "Crime in Public View" in P.J. Brantingham and P.L. Brantingham (eds.) *Environmental Criminology* (California: Sage Publications).

McDowell, L. (1989) "Towards an Understanding of the Gender Division of Urban Space", *Environment and Planning D: Society and Space*, 1: 59–72.

Newman, O. (1972) *Defensible Space — Crime Prevention through Urban Design* (London: Architectural Press).

Ong, J.H., Mak, L.F. and Chang, C.T. (1982) "Crime Prevention — Behaviour and Attitudes of the Public in Singapore", *Police Life Annual*, pp. 32–41.

Ooi, G.L. (1991) "Public Housing Policy and the Family — Case Study of Singapore", *International Journal of Housing Science and Its Application*, 15: 277–90.

Ooi, G.L. (1995) "Ang Mo Kio — The Makings of Place and Community in a New Town", in Yeoh, B.S.A. and Kong, L. (eds.) *Portraits of Places: History, Community and Identity in Singapore* (Singapore: Times Editions), pp. 158–79.

Peet, R. (1975) "The Geography of Crime — A Political Critique", *Professional Geographer*, 27: 277–80.

Peet, R. (1976) "Further Comment on the Geography of Crime", *Professional Geographer*, 28: 96–100.

Poyner, B. (1983) *Design against Crime — Beyond Defensible Space* (London: Butterworths).

Quah, S.R. and Quah, J.S.T. (1987) *Friends in Blue — The Police and the Public in Singapore* (Singapore: Oxford University Press).

Rand, G. (1979) "Multiple Housing Security through Environmental Design", in P.S. Hopf (ed.) *Handbook of Building Security, Planning and Design* (London: McGraw-Hill).

Report and Recommendations of the Public Housing Sub-Committee (1984) (Singapore: National Crime Prevention Council).

Riger, S. and Gordon, M.T. (1981) "The Fear of Rape — A Study in Social Control", *Journal of Social Issues*, 37: 71–93.

Salaff, J.W. (1988) *State and Family in Singapore: Restructring a Developing Society* (Ithaca: Cornell University Press).

Shaw, C.R. and McKay, H.D. (1969) *Juvenile Delinquency and Urban Areas* (Chicago: University of Chicago Press).

Smith, S.J. (1983) "Crime and the Structure of Social Relations",

Transactions of the Institute of British Geographers, 17: 427–40.

Smith, S.J. (1987) "Fear of Crime — Beyond a Geography of Deviance", *Progress in Human Geography*, 11: 1–23.

Smith, S.J. (1989) "Social Relations, Neighbourhood Structure, and the Fear of Crime in Britain", in Evans, D. and Herbert, D. (eds.) *The Geography of Crime* (London: Routledge), pp. 193–227.

Stollard, P. (1991) *Crime Prevention through Housing Design* (London: E & F N Spon).

Sutherland, E. (1939) *Principles of Criminology* (New York: Lippincott).

Tan, D.W.S. (1988) "The Strategy of Community Policing", *Police Life Annual*, pp. 59–66.

Teo, P. and Huang, S. (1996) "A Sense of Place in Public Housing — A Case Study of Pasir Ris, Singapore", *Habitat International* 20(2): 307–25.

Teo S.E. and Savage V.R. (1985) "Singapore Landscape — A Historical Overview of Housing Change", *Journal of Tropical Geography*, 6: 48–63.

Thrasher, F. (1928) *The Gang — A Study of 1303 Gangs in Chicago* (Chicago: University of Chicago Press).

Tng, B.H. (1992) "Beyond the Ten Years — The National Crime Prevention Council Takes a Bold Look at Strategies for the 90s", *Police Life Annual*, pp. 104–109.

Valentine, G. (1989) "The Geography of Women's Fear", *Area*, 21: 385–90.

Valentine, G. (1992) "Women's Fear and the Design of Public Spaces", *Built Environment*, 14: 288–303.

Warr, M. (1985) "Fear of Rape among Urban Women", *Social Problems*, 32: 238–50.

Wong, A.K. and Yeh, S.H.K. (eds.) (1985) *Housing a Nation — 25 Years of Public Housing in Singapore*, Maruzen Asia for Housing and Development Board, (Singapore).

Yeh, S.H.K. (ed.) (1975) *Public Housing in Singapore* (Singapore: Singapore University Press).

Yeoh, B.S.A. and Kong, L. (1994) "Reading Landscape Meanings — State Constructions and Lived Experiences in Singapore's Chinatown", *Habitat International*, 18: 17–35.

Yeow, P.L. (1993/94) *Through Women's Eyes — A Geography of Fear in Singapore*, Academic Exercise, Department of Geography, National University of Singapore, (Singapore).

Yin, P. (1980) "Fear of Crime among the Elderly — Some Issues

and Suggestions", *Social Problems*, 27: 492–504.
Yin, P. (1982) "Fear of Crime as a Problem for the Elderly", *Social Problems*, 30: 240–45.
Yip, L. and Ong, J. (1991) "How Crime Free is Crime Free Singapore". *Police Life Annual*, pp. 83-90.

Contributors

Associate Professor T. Balendra is a senior staff of the Department of Civil Engineering, National University of Singapore. He has overseen research and published widely on the subject of structural designs and earthquake tremors in the region.

Dr Myrna Blake is Senior Research Fellow at the Department of Social Work and Psychology at the National University of Singapore. She is greatly involved in work among the socially disadvantaged in Singapore and has recently published a book on the Eurasian community in Singapore.

Dr Kog Yue Choong is Executive Director at Beca Carter Hollings and Ferner (S.E. Asia) Pte Ltd, a major engineering design consultancy firm in Singapore. He has a Ph.D. from the National University of Singapore. He is also Adjunct Associate Professor at the Department of Civil Engineering, National University of Singapore. Currently, he is the President of the Association of Consulting Engineers Singapore.

Dr Kenson Kwok was Senior Research Officer at the Housing and Development Board before assuming his current position as Director of the Asian Civilisations Museum at the National Heritage Board. He is an architect and environmental psychologist as well and has a Ph.D. from the University of London.

Dr Lai Ah Eng is currently with National Archives Board, Singapore. She was Executive Research Officer at the Housing and Development Board before proceeding to do her Ph.D. at Cambridge University. She has recently published a book on ethnic relations based on a case-study of a public housing estate in Singapore.

Mr Liu Thai Ker was the Chief Executive Officer of Housing and Development Board before he assumed the post of Chief Executive Officer and Chief Planner at the Urban Redevelopment Authority in Singapore. Currently he is with RSP Architects, Planners and Engineers, (Pte) Ltd Singapore, a major architectural, planning and engineering design consultancy firm.

Associate Professor Linda Low teaches economics at the Business Policy Department at the Faculty of Business Administration, National University of Singapore. She has done a considerable amount of research and published widely in areas such as public finance analysis, privatisation, health financing and national development.

Dr Ooi Giok Ling is Senior Research Fellow at the Institute of Policy Studies. She has a Ph.D. from the Australian National University (Canberra). Her research and publications have focused on housing and urban studies, retailing, health care, the environment and she is currently on secondment to the Ministry of Home Affairs as Director of Research.

Dr Phang Sock Yong is Senior Lecturer at the Department of Economics and Statistics, and also Deputy Director of the Masters in Public Policy Programme at the National University of Singapore. She obtained her M.A. and Ph.D. from Harvard University. Her area of specialisation is Urban and Transportation Economics. She is the author of *Housing Markets and Urban Transportation* and *The Singapore Experience in Public Housing,* and has also published many academic journals.

Associate Professor Victor R. Savage, a Singaporean, did his undergraduate Arts (Honours) Degree at the University of Singapore and obtained his M.A. and Ph.D. in geography from the University of California, Berkeley. He completed his M.A. and Ph.D. under the Fulbright Hayes Scholarship. His research interest is mainly South-east Asia — historical and cultural landscapes, sustainable development and sustainable urban development issues. He is also interested in cultural landscapes and environmental issues on Singapore. Assoc Prof Savage is currently the Vice-Dean of the Faculty of Arts and Social Sciences, National University of Singapore (NUS).

Dr Sharon Siddique was the Deputy Director of the Institute of Southeast Asian Studies, Singapore. Trained as a development sociologist, she currently runs a consultancy firm which specialises in development studies in the South-east Asian region.

Dr Sim Loo Lee is Senior Lecturer at the School of Building and Estate Management. She has published widely on private property developments and public policy issues in Singapore.

Dr Brenda Yeoh is Senior Lecturer at the Department of Geography, National University of Singapore. She has a Ph.D. from Oxford University. She has written widely on cultural landscapes in Singapore and has an interest in gender studies.

Ms Yeow Pei Lin graduated with honours from the Department of Geography, National University of Singapore. She conducted the research she has contributed to this publication while an undergraduate.

Dr Lee Tsao Yuan

Director.

The Institute of Policy Studies

Tel : (65) 7792633

Fax (65) 7770700.

Internet ips@pacific.net.sg.